A Handbook on
ECONOMIC ENTOMOLOGY

The Author

Dr. Abhishek Shukla, *Ph.D.* is a Subject Matter Specialist (Plant Protection) and is working in Krishi Vigyan Kendra, Chomu, Jaipur, Rajasthan. He has more than seven years of experience in research and extension. He has more than thirty research papers and several popular articles to his credit. He has written two books in Hindi titled, *"Masalon mein Keet Prabandhan"* and *"Krishi Keet Vigyan: Ek Parichaya"*. He has been member of several scientific societies in India and abroad.

A Handbook on ECONOMIC ENTOMOLOGY

Dr. Abhishek Shukla
Subject Matter Specialist (Plant Protection)
Krishi Vigyan Kendra, Chomu – 303 702
Jaipur, Rajasthan, India

2025
Daya Publishing House®
A Division of
Astral International Pvt. Ltd.
New Delhi – 110 002

First Impression, 2009
Reprinted, 2025
ISBN 978-81-7035-615-8 (HB)

Published by : **Daya Publishing House®**
A Division of
Astral International Pvt. Ltd.
– ISO 9001:2015 Certified Company –
4736/23, Ansari Road, Darya Ganj New
Delhi-110 002
Ph. 011-43549197, 23278134
E-mail: info@astralint.com
Website: www.astralint.com

Preface

Insects are always considered as harmful by mankind. But it is quite interesting to note that there are various insects which are of economic importance and provide various useful products like honey, silk and lac. Besides there commercial value, they are important as many of them are serious pests of crops. This book *A Handbook on Economic Entomology*, covers selected subjects are part the syllabi of various Agricultural Universities/universities in India.

Nowadays, there is great emphasis on development of applied science. Various Government and Non Government agencies are keenly in promoting this branch of science. Apiculture and Sericulture have already established their great economical and commercial values. Lac culture is also progressing towards its goal.

There was a long need for a simple and comparative book on this very important aspect of Entomology. My prime objective is to see that the book helps the students and practitioners to learn and practice the various aspects of economic entomology. I have tried hard to collect and compile useful information about the subject for the students and practitioners.

I owe gratitude to Dr. Ashok Kumar, Head, Department of Entomology, Maharana Pratap University of Agriculture and Technology, Udaipur for his valuable suggestions and guidance during the course of preparation of the manuscript. I am heartily thankful to Dr. S. S. Rathore, Programme Coordinator, Krishi Vigyan Kendra, Chomu, Jaipur and Dr. Sushil Kumar Sharma, SMS (Soil Science), Krishi Vigyan Kendra, Chomu, Jaipur for their constant encouragement which helped me in developing a confidence which was much needed and is gratefully acknowledged.

Lastly I shall be glad to receive constructive comments, suggestions, and corrections from seniors, colleagues and students.

Dr. Abhishek Shukla

Preface

Contents

1

Introduction

Agricultural crops have been prone to damage due to insect pests since time immemorial. Off and on these pests have caused colossal crop losses. The locust plague during 1925-31 is a classical example to total destruction of crops in Western parts of India. Considering the role of insects in agriculture, they can be mainly classified into three major group's *viz.*, harmless, harmful and useful insects. The first group *i.e.*, harmless insects, includes thousands of insect species which does not affect the economy of man. The remaining two groups considered important as they affect economic interest of man either by useful or destructive activities.

Usefulness of Insects in Agriculture

Some insect species are useful for mankind. There advantages are as follow:

I. Direct Advantages

There are some insects which are of great economic importance as they produce valuable things like honey, silk, lac, dye, etc. These insects are called as beneficial insects.

II. Indirect Advantages

(*a*) Pollination

Honey bees and some other wasps and insects pollinate the crops and increase the yield 10-80 per cent.

(*b*) Parasite and Predators

Some insects feeds on harmful insects and reduce their numbers considerably. They are either parasite or predators. *Trichogramma* is a classical example of parasitoid while lady bird beetles, mantis are some of the effective predators.

(c) Source of Drugs

Some specific drugs are prepared from the insects *e.g.* powdered dry cockroaches are said to be useful in urine trouble, bleaster beetles are good source of cantharidine which are useful in removing sterility in cattle.

(d) Scavengers

There are some insects, which act as scavengers by way to eating waste materials, *e.g.* flies.

(e) Weed Killers

Some insect feed upon harmful weeds and are useful in biological control programme of weeds, *e.g. Mexican beetle, Zygromma* controls *Parthanium* weed effectively.

(f) Insect as Food

Some insects are used in the form of a food, *e.g.* queen termites, locusts and larvae, in different parts of world.

Losses Caused by Insects to Agriculture

The insect caused damage to crops by two ways *i.e.*, direct or indirect losses.

I. Direct Losses to Crop Plants

When every plant part is affected or damaged by insect. The direct losses to different plant parts are as follow:

(a) Root

The soil inhibiting insects cause considerable losses to plant roots *e.g.*, termite, white grub, etc. Some of them completely destroy the roots resulting in the death of plants, where as some of them eat only the portion of the roots due to which plant becomes weak and sickly in appearance.

(b) Stem

Many insects and their immature stages enter into the stem of various plants and cause serious damage, *e.g.* maize stem borer, *Chilo partellus*.

(b) Leaf

Leaf feeding insects cause damage by different ways:

1. *Leaf eaters*: The insects having biting and chewing mouthparts and feed on the leaves. *e.g.* grasshopper, locust, beetles and caterpillars.
2. *Leaf miner*: They live between the upper and lower epidermis of leaves and feed green portion. *e.g.* Pea leaf miner, *Phytomyza atricornis*.
3. *Leaf roller*: Some of the larvae feed plant leaves and they also roll them. Due to this leaves shrivel and fall. *e.g.* cotton leaf roller, *Sylepta derogates*.
4. *Sap feeders*: Many insects having piercing and sucking mouthparts *e.g.* aphids, jassids, whiteflies, thrips, etc. They drain the sap from leaves and due to continuous sap sucking leaves become brown or yellow and fall.

(*d*) Flower and Fruit Feeder

Many beetles, thrips and jassids feed on various parts of flowers, while many feed on immature fruits by sucking fruit juices or bore the fruits, *e.g.* fruit sucking moth and fruit flies.

(*e*) Seed Feeders

Some insects feed either on the developing seeds or the immature seeds in plants. Gram pod borer, *Helicoverpa armigera* feeds on immature seeds of sunflower.

II. Indirect Losses to the Plants

Besides causing direct losses to the plants there exist some insects species which produce honey dew, results in the formation of black mould. The black mould interrupt in the photosynthesis and respiration of the plants and finally adversely affect plant growth. Number of insects are known which transmit viral diseases in plants. Whiteflies is the most prominent group which is responsible to transmit various viral diseases in the plants.

2

Apiculture

Man has at several occasions proved his intelligence and superior skill by converting the wonderous capabilities of the animal world to his material advantage. One of the finest discoveries in this line is our knowledge regarding the procurement of honey collected by bees. The use of honey was known to man since the earliest of times in India as well as in other parts of the world, because it is a nourishment of high food values. In olden days, and in many places still, the methods utilized for obtaining honey from the hive is very crude. Crushing of hive to extract honey destroys the endeavouring efforts of bees which they serve during the formation of comb and in providing nourishment to the developing larvae. Not only are the comb and larvae destroyed, this method also gives very unhygienic honey. Modern investigators have overcomed these difficulties by using scientific methods of bee keeping which is known as apiculture.

Bee keeping has been practiced in India since time immemorial. The earliest references date back to Vedas and Ramayana. But scientific bee keeping with the help of simple machine and untouched by hand is a comparatively new venture. The western method of frame-hive was first introduced in Bengal in 1882 and in the Punjab in 1883-84. In recent times progress has been rapid, for practical apiculture is an art which has, during the last forty years, not only undergone a complete revolution but has attained a development and multiplicity of detail. It was due to the efforts of the khadi and village industries that bee-keeping acquired its present co-ordinated national status in India. In 1962 the commission under its directorate of bee-keeping established the Central Bee Research Training Institute at Pune. This institute is gradually progressing and has made its sub-stations at Kodaikanal, Mercara, Castle Rock, Mahabaleshwar, Kangra and Kashmir.

Types of Honey Bee

Honey bee belongs to the class Insecta, order hymenoptera and family apidae. There are four well recognized types of bees found in the world.

1. *Apis dorsata* (Rock bee)
2. *Apis florea* (Little bee)
3. *Apis indica* (Indian bee)
4. *Apis mellifica* (European bee)

Out of these four types, three are common in India. They are *Apis dorsata*, *Apis florea* and *Apis indica*.

Apis dorsata, which is commonly called as rock bee, is the largest Indian variety with an average size of about 20 m.m. It builds large comb (0.90 x 15 metres) on tree branches, under caves, or under roofs of high buildings. They are migratory species as during June and July they swarm to the hills, but in winter come back to the plains. They have been found up to the height of 3,500 feet above sea level. This variety is yet to be successfully hived. Researches are going on, on the behaviour of this variety in order to domesticate them. This variety has the highest honey yield (average 15 kg. per colony per year) amongst Indian bees. Sometimes, the yield exceeds 30 kg. per colony per year. This bee is notorious for its ferocity and tendency to make unprovoked, sometimes fatal, mass attack on persons who approach its hive.

Apis florae, which is commonly called as little bee, is a miniature of the rock bee. It is a plain species and rarely occurs above 1000 feet of sea level. It builds small comb (about 15.24 cms. Across) on the branches of trees, or in bushes, or under the wall of the buildings. The yield of honey from this type is very little (few ounces per colony per year), and the production does not compensate the labour undergone on it.

Apis-indica, popularly known as Indian bee, is of commonest occurrence on the plains and forests of India. There are several regional strains of it, of which plain, transitional and hill varieties are three recognized types. *Picea* strain is found in hills at an altitude up to 7,000 ft, *Pironi* is distributed along the transition between an altitude of 3,000 to 4,000 ft, whereas, *Lighter indica* is a plain strain found up to an altitude of 1,500 ft. It builds several parallel combs (about 30 cms across) in protected places like hollow of trees, caves, in rocks and in other such cavities. Due to their mild nature and average output of honey between 3 kg. to 5 kg. per colony per year, they are amongst the best of the Indian variety to be hived in artificial conditions.

Apis mellifica or European bee is very common all over the Europe. This bee is similar to *Apis indica* in its habitates. There are several varieties and strains of this bee amongst them the Italian variety is the best. It yields an average of honey per year per colony. Attempts to domesticate this be in India or large scale has yet not been proved to be a success.

The yield of honey from Indian bee is quite poor compared to their Italian or South African counterparts. The South African yield of 100 kg. per hive is about twelve times more than the Indian average of 4.5 kg. per hive per year.

Castes of Honey Bee

Honey bee is a social insect. The nest of the honey bee is known as the bee-hive. A hive in summer consists of 32 to 50 thousand individuals, depending on the locality. A colony is termed 'weak' or 'strong' according to the number of worker bees it possesses. There are three types of individuals in a colony, namely the Queen, worker and drone. Due to the existence of several morphological forms, bees are said to be a polymorphic species. All these three castes depend on each other for their existence.

Queen

It is a diploid, fertile female. The presence of queen in a colony is a must. The size of the body of queen is much larger than other castes of bees of the colony. Her legs are strong, she is always walking about on the comb. The queen has a sting curved like a scimitarat the tip of the abdomen, which is in fact a modification of the egg-laying organ known as ovipositor. The sting serves as an organ of defence. She never uses it against anybody except her own caste. The queen is responsible for laying eggs for a colony. She lays about 1000 to 1500 eggs everyday and lives a life of two three years. However, the number of eggs laid per day may vary from individual to individual, and it has been found that a queen may produce as much as 6,000 eggs per day. She lays both fertilized eggs (from which females develop) and unfertilized eggs (from which males develop).

Worker

It is a diploid, sterile female. The size of a worker is the smallest among the castes but they constitute the majority of the bees in a colony. Their function is to collect honey, to look after youngones, to clean the comb, to defend the hive and to maintain the temperature of the hive. Numerous adaptations have occurred in the worker for performing her various functions. The body is covered with branched hairs so that when a bee visits a flower, pollen grains adhere to the hairs and other parts of the body. The worker cleans off pollen grains with special structures the antenna cleaners on ech foreleg, pollen brushes on all legs and pollen combs on hind legs. All pollen

Drone Queen Worker

Figure 2.1: Casts of Honey Bee

are stored in the pollen basket present on the outer surface of tibiae on hind legs. Water and nectar are gathered by means of sucking mouthparts which are modifications of the maxillae and labium.

Workers are provided by a sting at the tip of the abdomen which is a modified ovipositor. A large poison storage sac is connected with the base of the sting. Two acidic and one alkaline gland mix their secretion to form poison which is injected by the operation of muscles to other animals. During the withdrawal from the prey's body, the sting along with other poison apparatus are turn off, resulting in the death of that particular bee. Workers are female but are incapable of producing eggs. The life span of a worker bee is 4-5 months but during hard working days they persists for five to six weeks only.

Drone

It is haploid, fertile male. The males are larger than workers and are quite noisy. They are unable to gather food, but eat voraciously. They are stingless and their sole function is to fertilize the female (queen). The number of drones in a colony varies from 200-300, but during bad season they are driven out. The drone develops parthenogenetically from unfertilized eggs.

Life Cycle of Honey Bee

To become an ideal apiculturist, it is necessary to know the habit and life history of bees. Each hive contains one queen, several drones and innumerable workers. During nuptial flight, many drones follow the queen. The drone which successfully copulates with queen looses its copulatory apparatus and ultimately dies. The sperms get stored in the spermathecae of the queen which after returning to the hive start laying eggs. The number of eggs laid may exceed 2,000 per day but it depends largely upon the temperature and availability of food. A queen can lay both fertilized and unfertilized eggs. The entire process of egg laying is believed to be under voluntary control of the queen.

Eggs are small pearly white and spindle shaped. Unfertilized eggs are laid in much spacious drone cells where they develop into drones. Larvae hatchout from the eggs after about three days of egg laying. The unfertilized eggs develop parthenogenetically into male and fertilized eggs develop into females. During first 2-3 days all larvae are fed on a special food the 'Royal Jelly' secreted by the pharyngeal glands of the young workers. After that coarser food, the 'Bee Bread', which is a mixture of honey and pollen grain is given. However, the queen forming larvae are fed on royal jelly for full larval life and they are taken for further development into a special chamber called the queen's chamber. Differential food types results into difference in size of the developing larvae. The larvae moults several times and on the seventh day enter the quiescent period of pupation. Now the worker bees cover the cell with a thin layer of wax. Inside the cell larva spins a delicate silken cocoon around itself and turns into a pupa. During pupal stage, the legless white larva undergoes startling changes not only of external form but also internal organs. Finally the adult comes out by cutting the wall of cocoon first and secondly by breaking the wax covering. Time taken by queen, worker and drone for their development varies. It is given as below:

	Egg	*Larva*	*Pupa*	*Total*
Queen	3 days	5-1/2 days	7-1/2 days	16 days
Worker	3 days	6 days	12 days	21 days
Drone	3 days	6-1/2 days	14-1/2 days	24 days

Worker bees after coming out of the pupal case starts working and their duties changes with the advancing age. Collection of nectar and pollen from flowers is done in the last phase of their life, *i.e.* after 25th day of their birth.

Regarding sterlity of worker bees, which are females, there are two views. One says that the differential type of food given to larvae lead the degeneration of body size and reproductive organs. Other says that the mandibular gland of the queen secrete a "Queen substance" containing 9-oxodetrans-2-enoic acid which is sprayed upon the whole body by the queen. The workers lick the queen substance from her body, which results in the inhibition of the activities of ovary. It has been seen that if the workers are kept away from queen, their ovary ripens.

Structure of a Hive

The highest degree of nest construction among insects is found in bees. The architecture of the nest is unsurpassed and unparalleled in the animal kingdom. The hive and comb of the bees are formed mainly by workers. A comb is a vertical sheet of wax, composed of a double layer of hexagonal cells projecting in both directions from central wax-sheet.

Comb hangs vertically downward, while cells are horizontal in position. The hexagonal shape of cells accumulates maximum space in minimum use of wax and labour. The wax used in building of a comb is secreted from the wax glands present in the abdomen of worker bees. This wax has the highest melting point *i.e.*, 140ºF. Before use, the wax is masticated and mixed with secretions of the cephalic glands to convert into a plastic substance. The resinous substance called "propolis", prepared from pollen, is used in making the comb water-proof, and it also helps in filling the cracks and crevices in the hive.

The cells of the comb are of various types. The 'storage cells', which contains honey and pollen are generally built on the margin and at the top of the comb. The 'brood cells', which contains the young stages are built in the center and the lower part of the comb. Brood chamber is further divided into three types, namely Worker-chamber, where developing workers are reared; Drone-chamber, where developing drones are reared and the Queen-chamber, which is larger than other and where the larvae developing into queens are reared. There is no special chamber for adults. They move on the surface of the comb.

Scientific method of bee-keeping has been developed after the extensive studies of bee behaviour, their way of functioning and their mode of reproduction.

Essentials for Starting a Bee Hive

A beginner should learn in detail about the habit and behaviour of the bees. For this it is advisable to join the experienced bee keepers of that area.

To procure the hive and other tools connected with bee-keeping. The khadi and village industries commission are providing all sorts of assistance including monetary grant and technical knowhow.

The hive should be placed in a locality rich in vegetation especially the flowering plant. If several hives are kept in a plot, the distance between two hives should be atleast six feet.

The hive should face East. It should receive sunlight during morning and evening and some shade during mid-day.

Water should be available nearby and an open space infront of hive entrance is necessary.

It is advisable and economical to collect bees from a particular locality.

Spring is the best season for starting bee-keeping as during this season swarming occurs and bees can be easily procured to be lived.

Bee Hive and Other Tools Connected With Beekeeping

Bee Hive

The modern bee hive is based on certain principles and is called "movable frame hive". Hive is made up of wooden box. The box may be single walled or double walled. Single walled hive is light and cheap so it is more common, whereas the double walled hive is costly and heavy, it is however, more durable and provides better protection to the bees. The insulation provided by a single walled hive is not enough to tolerate the fluctuations of atmospheric changes outside. Double walled hive with sufficient insulation keeps the hive warm in winter and cool in summer.

Modern hive has a basal plate or bottom board on which is placed a wood box called the "Brood chamber" A bottom pore in brood chamber act as bees entrance. Inside the brood chamber has several frames hanging vertically from the top. These fames can be removed independently, that's why the modern hive is called as movable frame hive. The distance between two frames is known as "bee space". This space serves as a passage for the movement of the bees and at the sametime it is very important because it is the space in which the bee should not form a comb. If the bee space is kept less than the correct size, the bees join up the two combs and if the space is more than the correct size, bee forms an independent comb in between two frames. In both cases it is impossible to remove the frames independently. There is another similar chamber above the brood chamber called as "super". This chamber is meant for the storage of honey only and here queen is never allowed to enter. To prevent the entrance of queen into super, queen excluder is used between brood and super chamber. On the top of the super there is an inner covering and then a roof.

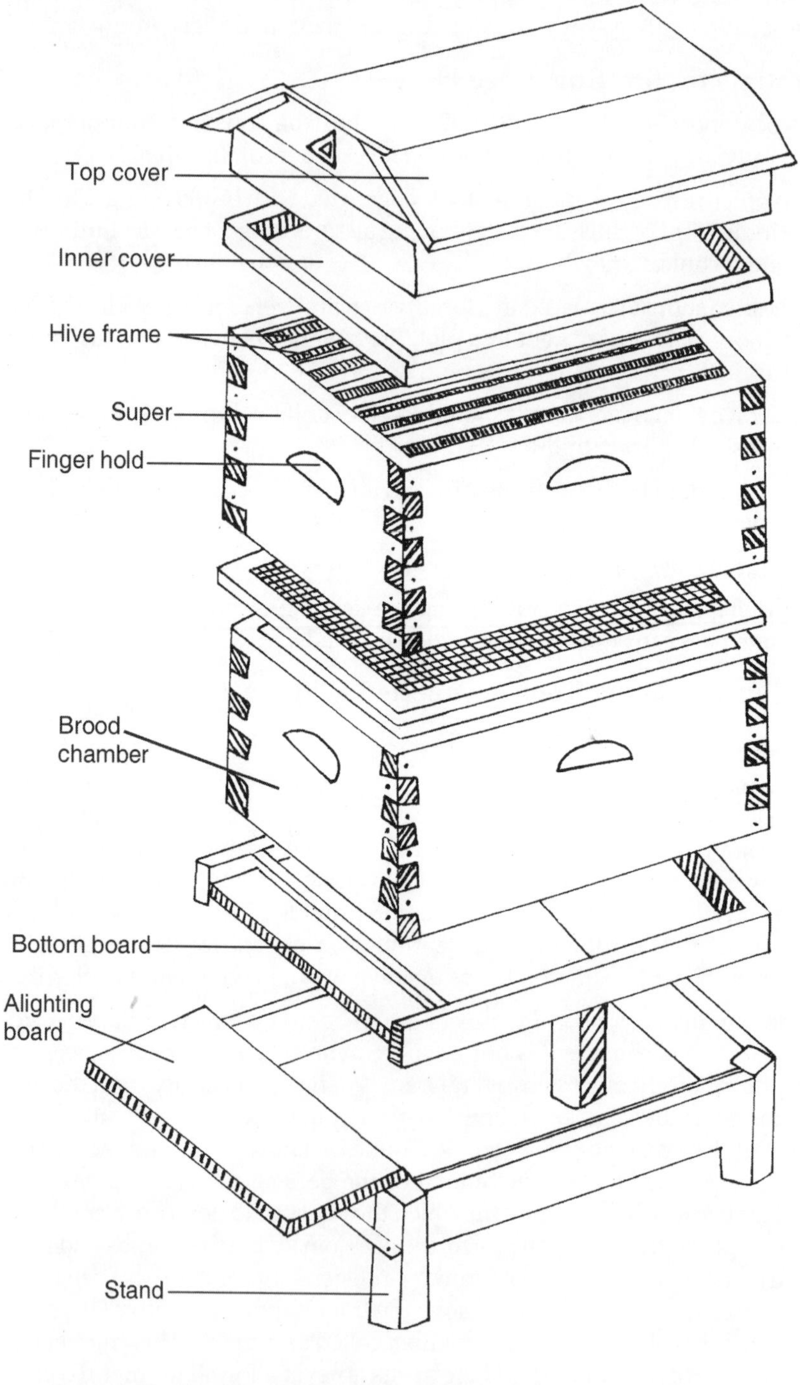

Figure 2.2: Different Parts of Bee Hive (Langstang Frame Hive)

(A) Bee space (Black colour)

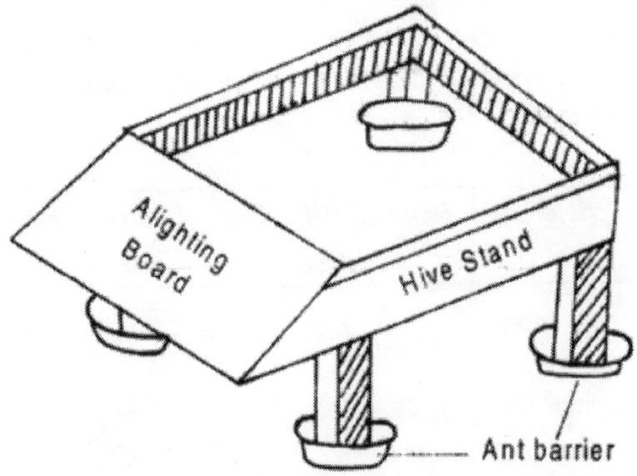

(B) Hive stand showing ant barrier

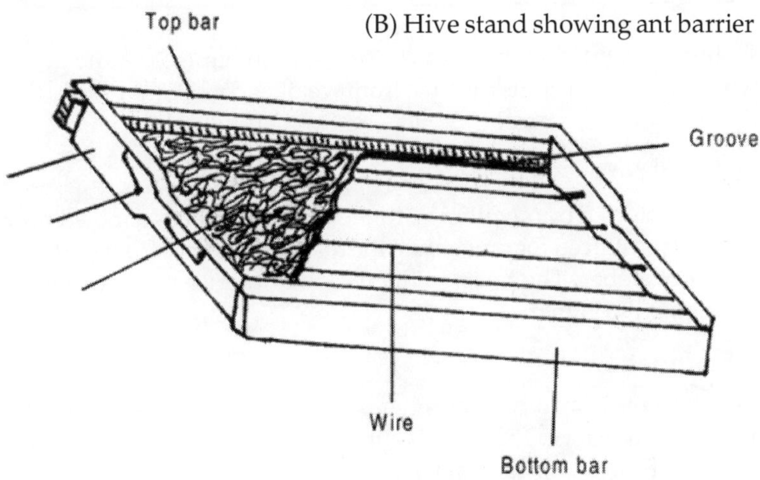

(C) Hoffman self spacing type hive frame

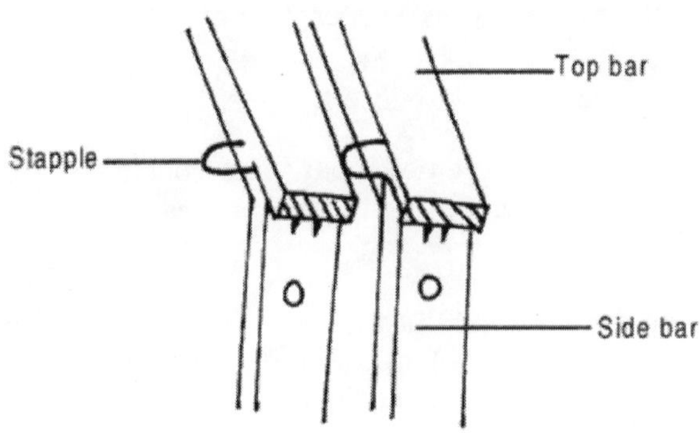

(D) Staple-space

Figure 2.3: Different Parts of Bee Hive

Different types of hive are being used in different parts of the world. In India generally, three types of bee-hive namely Langstroth, Newton and Jeolikote are in practice. The detail of the size and number of the frames used in different types of hive are given in tabular form below:

Type of Hive	*Number of Frames*		*Size of Frames*		*Recommendation*
	Brood	*Super*	*Brood*	*Super*	
Newton	7	7	8"x5$^1/_2$"	8"x2$^1/_2$"	It should preferably be used in the plains
Jeolikote	8	8	12"x7"	12"x3$^1/_2$"	-do-
Langstroth	11	11	17$^5/_8$"x9$^1/_8$"	17$^5/_8$"x9$^1/_8$"	It should preferably be used in hill region

The hive is painted with two coats of white colour (sometime green or yellow also), which not only protects the hive from weather conditions but also help the bees to easily recognize their hive.

Queen Excluder

This consists of a frame fitted with metallic wire net assembled together 0.150 inches apart. It is utilized for preventing the queen's entrance from the brood chamber to the super chamber. The holes in the net do not cause any inconvenience to the workers to pass through it.

Comb Foundation

It is a sheet of bee wax on both sides of which exact shape of different cells of the comb is made in advance. The sheet is cut to the size of hanging frames and are fitted inside it. It can be used for several years.

Bee Gloves

They are leather gloves used by bee keepers to protect heir hands from the sting of the bees.

Bee Veil

A bee veil is a covering to protect the keeper's face from the sting of the bees. It is made up of fine net, usually silken through which bees cannot pass.

Smoker

It is device used by keepers to subdue the bees if irritated during hive inspection. The materials used for producing smoke are anything like rotten wood, chips, wood latches, waste papers etc. In many parts of the world carbolic acid is used as bee quiter in place of smokers. It is powerful antiseptic, a good repellent and is used in diluted form.

Hive Tool

It is a flat, narrow and long piece of iron which help in scraping the dirty materials deposited by the bees especially bee glue and superfluous pieces of comb on the inner walls of the hive.

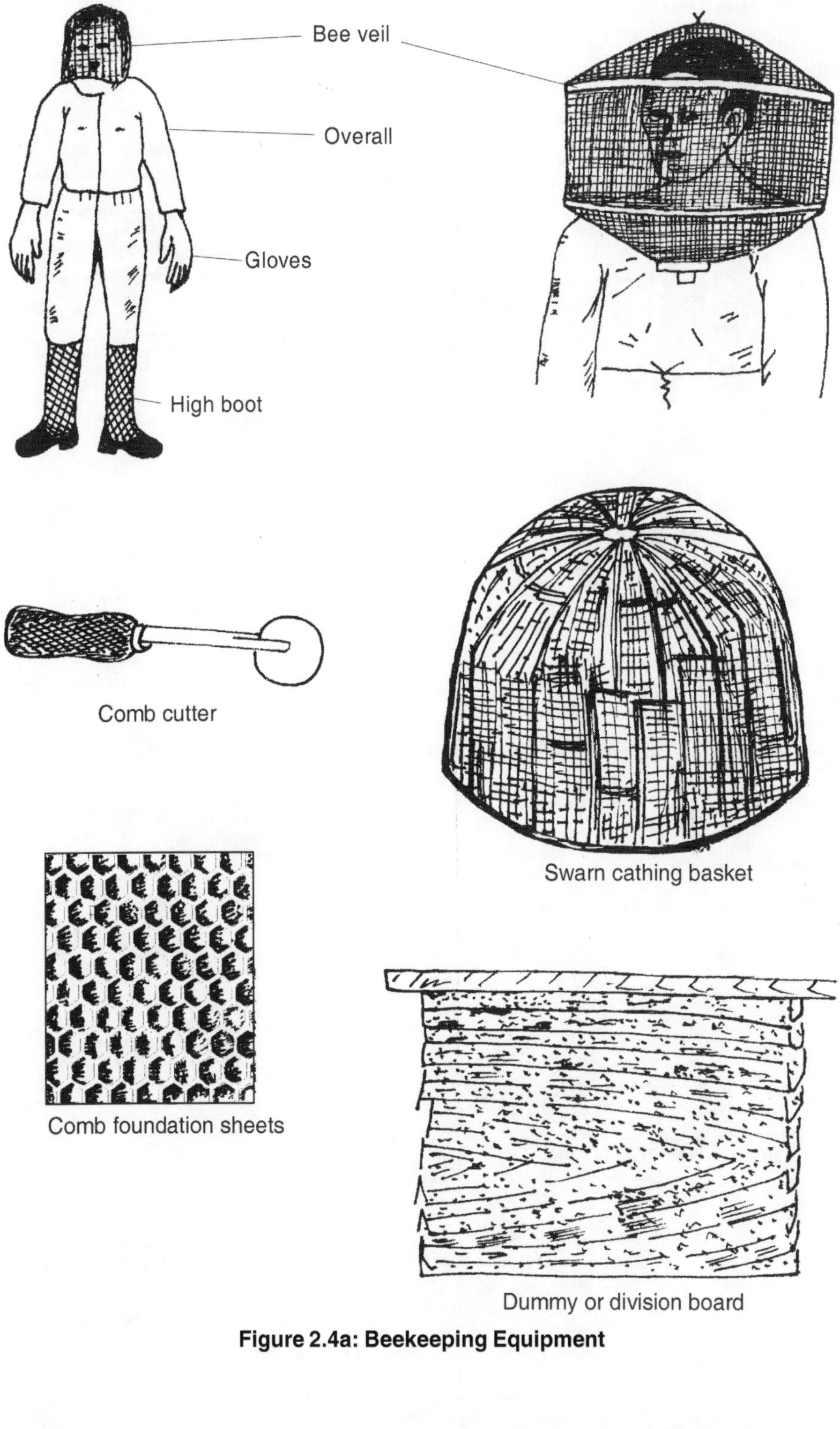

Figure 2.4a: Beekeeping Equipment

Uncapping Knife

It is a long, broad iron piece which helps in removing the cap of the hive for inspection at regular intervals.

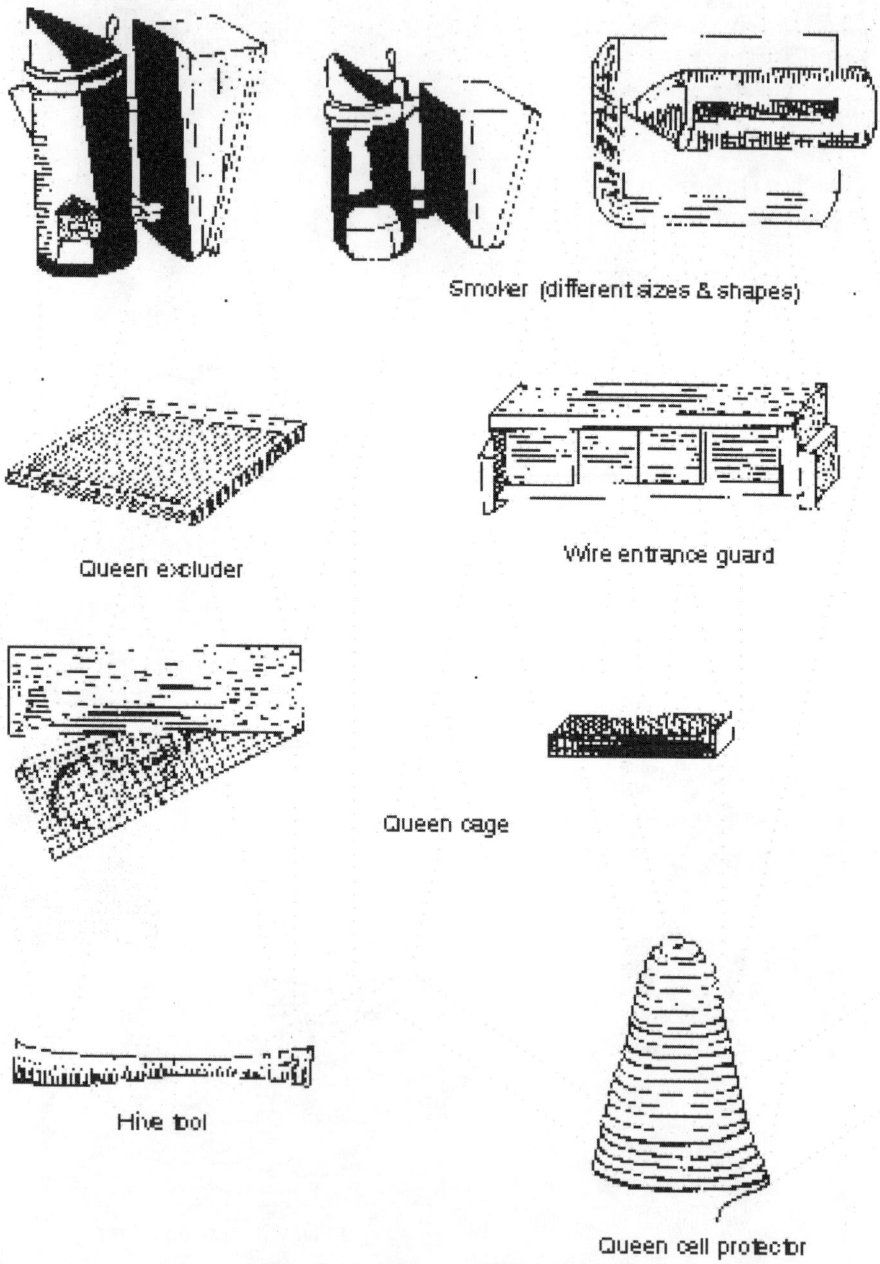

Figure 2.4b: Beekeeping Equipment

Bee Brush

It is a large brush often employed to brush off bees from honey combs particularly at the time of honey extraction.

Queen Introducing Cage

It is a pipe made up of wire nets through which the queen cannot pass. Both the ends of the pipe are opened. This tool is used for keeping the queen arrested in the hive for about twenty four hours, so that she gets acquainted with the hive as well as the worker bees. After putting the queen in the cage both the ends of the pipe is closed with "queen candy" (made by kneading fine powdered sugar with a little honey). Now the cage is placed inside the hive. In about twenty four hours time a hole is made in candy due to eating of candy by workers as well as queen, thereby releasing the queen. If the bees fail to eat the candy within 48 hours, the queen should be released directly.

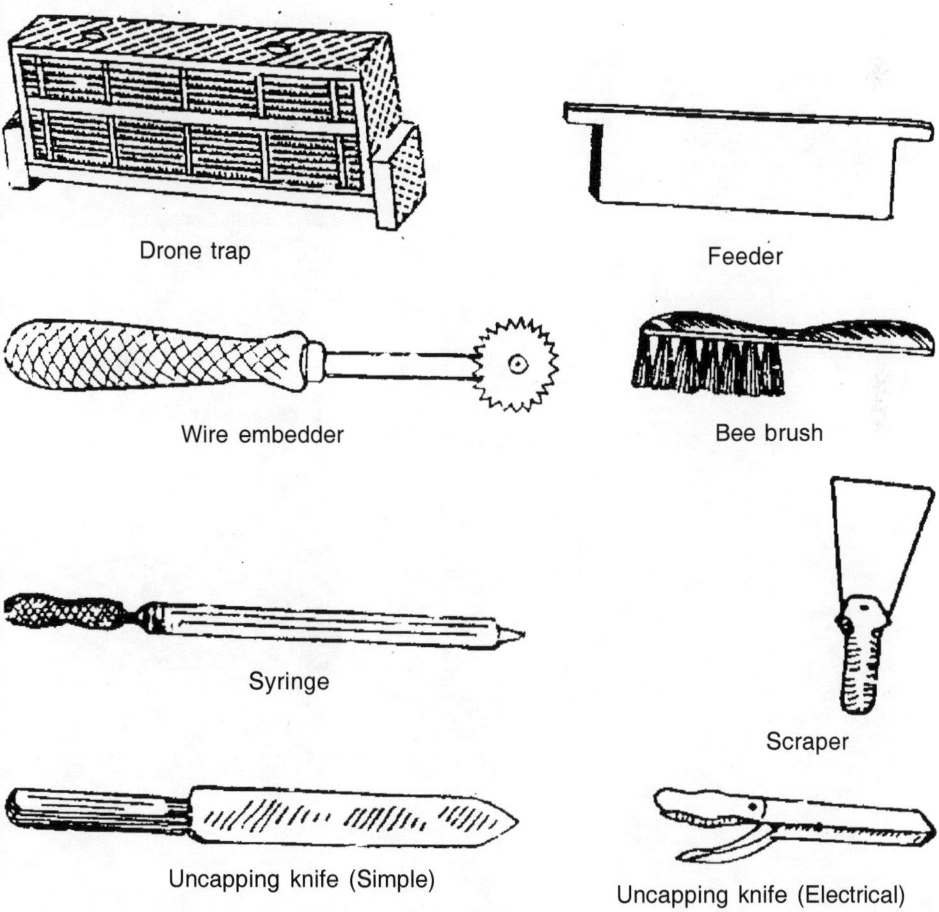

Figure 2.4c: Beekeeping Equipments

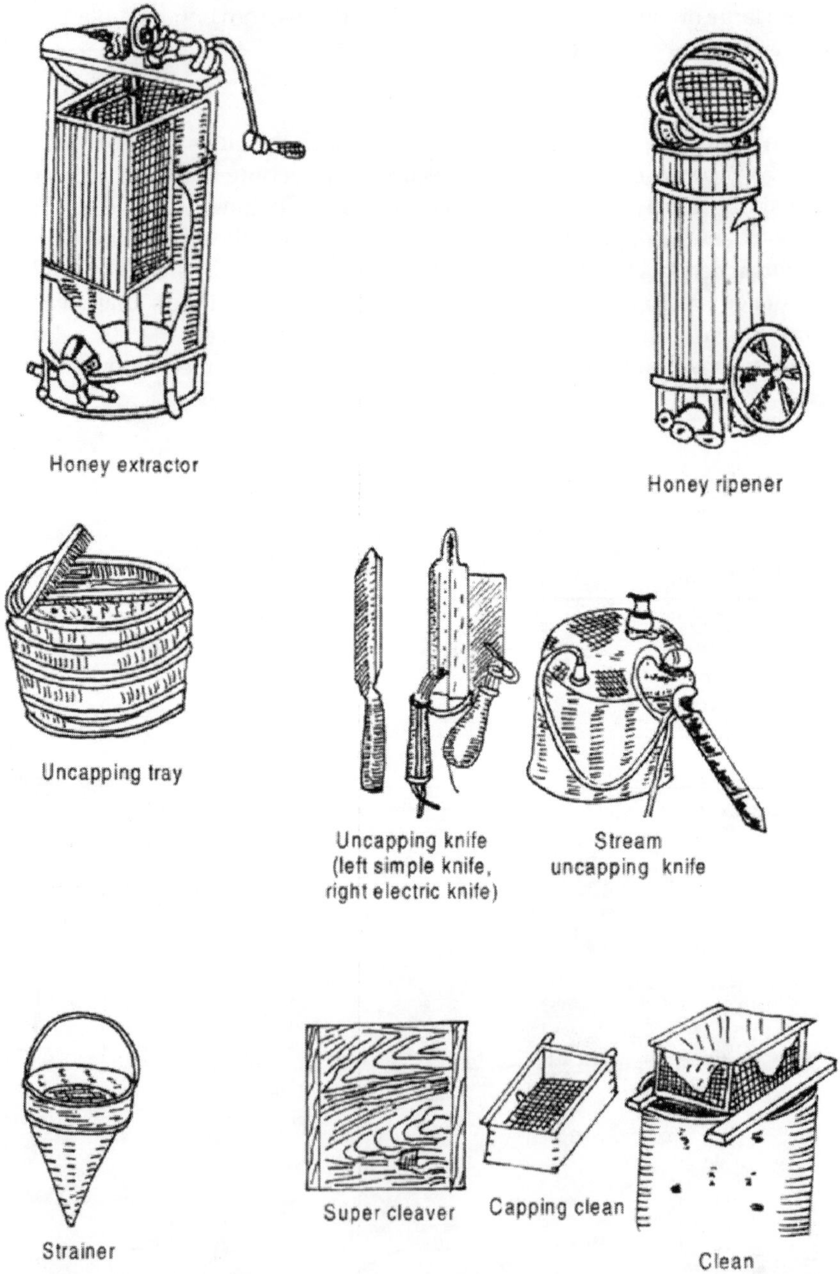

Figure 2.4d: Beekeeping Equipments

Feeder

During drought and lack of natural food bees are fed with artificial food. Sugar syrup taken in a basin is placed over the frames of the brood chamber. For preventing the bees from sinking in the syrup, few green grass blades are placed in the basin along with the sugar syrup.

Honey Extractor

This instrument is used for extracting honey from the frames without any destruction to the comb. It is a drum made up of metal, having several pockets around a rotating wheel. The frames are made to hang from these pockets and the pockets are made to rotate round a central axis. This rotation creates a centrifugal force which separates the honey from the comb. The collected honey is taken out from the drum through a hole at the bottom. After extraction of the honey, the combs and frames are utilized again.

Hive Entrance Guard

During swarming season a device similar to queen excluder is placed in front of the hive entrance which prevents the escape of queen.

After collecting the hive and various tools mentioned above, as apiculturist comes in a position to start the actual process of bee keeping. The first step in this regard is procurement of bees and their hiving in a particular place.

For collection of bees, spring is considered the best season in India, as swarming occurs in this season. A swarm can be collected in cluster from the branches of trees, from corners of the houses or from cavities. From such places bees are shaken into cloth bags and immediately the mouth of the bag is closed. The collected bees in the afternoon is released in particular hive. Before releasing the bees the queen is placed in the hive in a "queen introducing cage". For quicker relief sugar syrup as artificial diet may be given to the bees. Within 24 hours the queen and workers get familiar to each other as well as with the hive. In this way they establishes a colony. For a few days, hive entrance guard is used to prevent the escape of queen.

Few Important Points Regarding Handling of Bees

1. The hive should not be opened too often. It can however, be opened twice or thrice a month. The time for examining the hive in summer is morning and evening whereas it is the warmer part of the day in winter.
2. There should be a constant watch to place additional chamber at the time when bees want to expand their nest. Such expansion may occur during the peak flowering season.
3. A bee keeper should keep an eye on the bees to see if they are being attacked by wax moth.
4. During swarming season the bee keeper should be on the alert and should keep an eye on the behaviour of the bees.

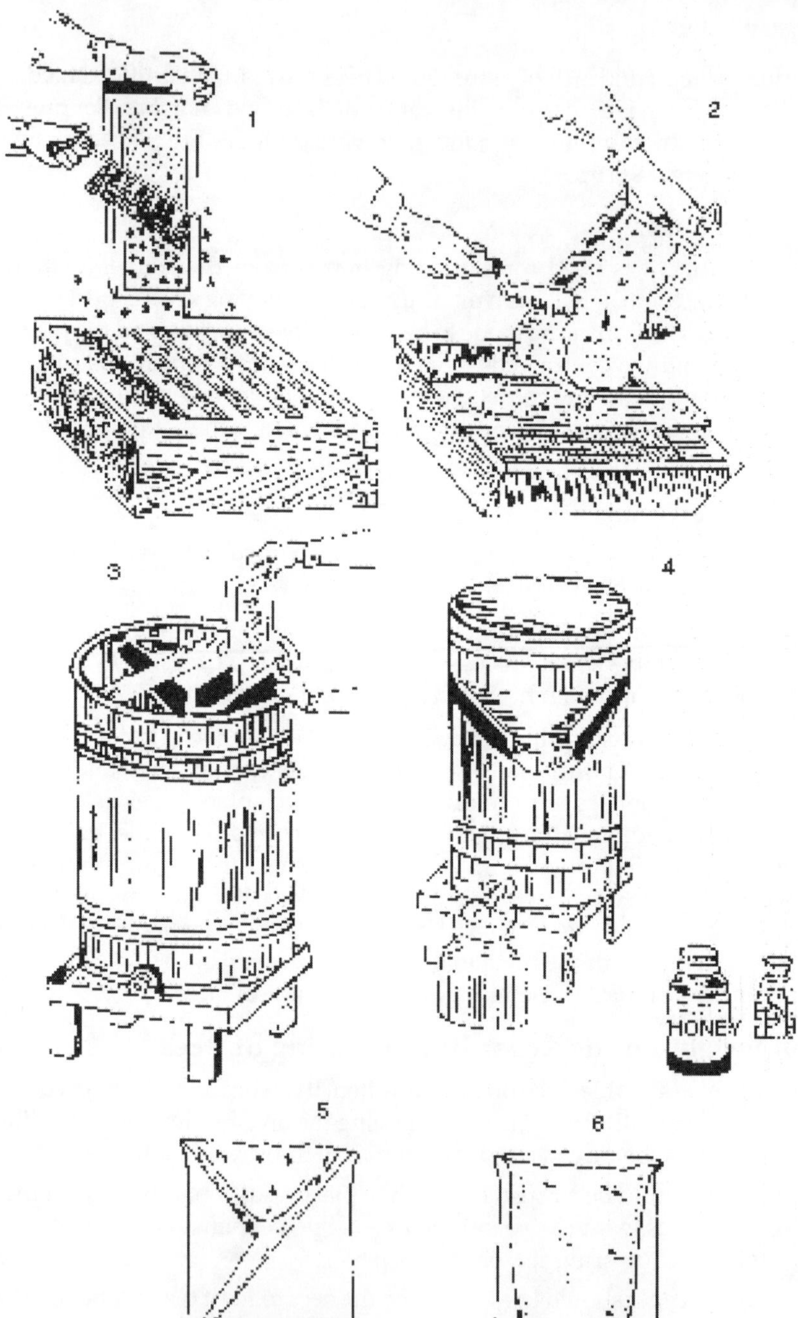

Figure 2.5: Honey Extraction

Swarming

It is a natural phenomenon whereby mass movement of animals from one place to another takes place. Bees also undergo swarming especially during spring season. All the bees comes out of the hive, they get divided into two or three colonies and move in different directions. Queen and drones also follow them. In case the queen does not leave the hive, there are chances that bees may return back to their hive.

- ☆ During swarming season following precautions should be taken in advance.
- ☆ Ample space for expansion of nest should be provided.
- ☆ There should be sufficient ventilation in hive.
- ☆ Hive should always have a young and vigorous queen.
- ☆ During swarming season queen's wing should be clipped.

If all these measures fail then bees should be transferred to a different hive which give them a chance for survival.

Bee Dance

Insects by their mysterious behaviour have at several occasion astonished human beings but the honey bee has surpassed all of them by their peculiar movements, better known as "bee dance". Bee dance is not only a device which provides a clue to the feeding place, but it also communicates the exact location, the strength and the distance of the feeding place from the hive. This peculiar sense of understanding and communication among bees has led Austrian born zoologist Carl Von Frisch to be honoured with the Noble prize in the year, 1973.

All the bees of a hive donot go out for search of food. There are certain workers generally called as "foragers or scout bees" who perform this duty and on their information other workers move to the feeding place. The forage after discovering a suitable source of food returns back to the hive, loaded with nectar or pollen and inform other inmates about her find by performing the dance. The dance is performed in two ways, round dance and tail wagging dance. Food sources less than hundred metres from the hive is indicated by dance in circular fashion whereas food at a distant of more than hundred metres is indicated by a dance.

Round dance convey the information about the source of food only, but tail wagging dance indicates distance, direction as well as the quantity of the food. Between each semicircle, the bee moves along the vertical plane waggling its abdomen side to side. The distance of food source from the hive has a definite proportional relation to the number of wagges. The frequency of wagging decreases with the increase in distance. The performance is so accurate that one can calculate the distance of food source by counting the number of wagges or by recording the time taken in dance through a stop watch. The movement of sun has a definite relation with the direction of dance. At a particular hour of the day of forager makes the same angle from the vertical plane during her waggling dance which at that hour stand among the triangle of hive, food and sun.

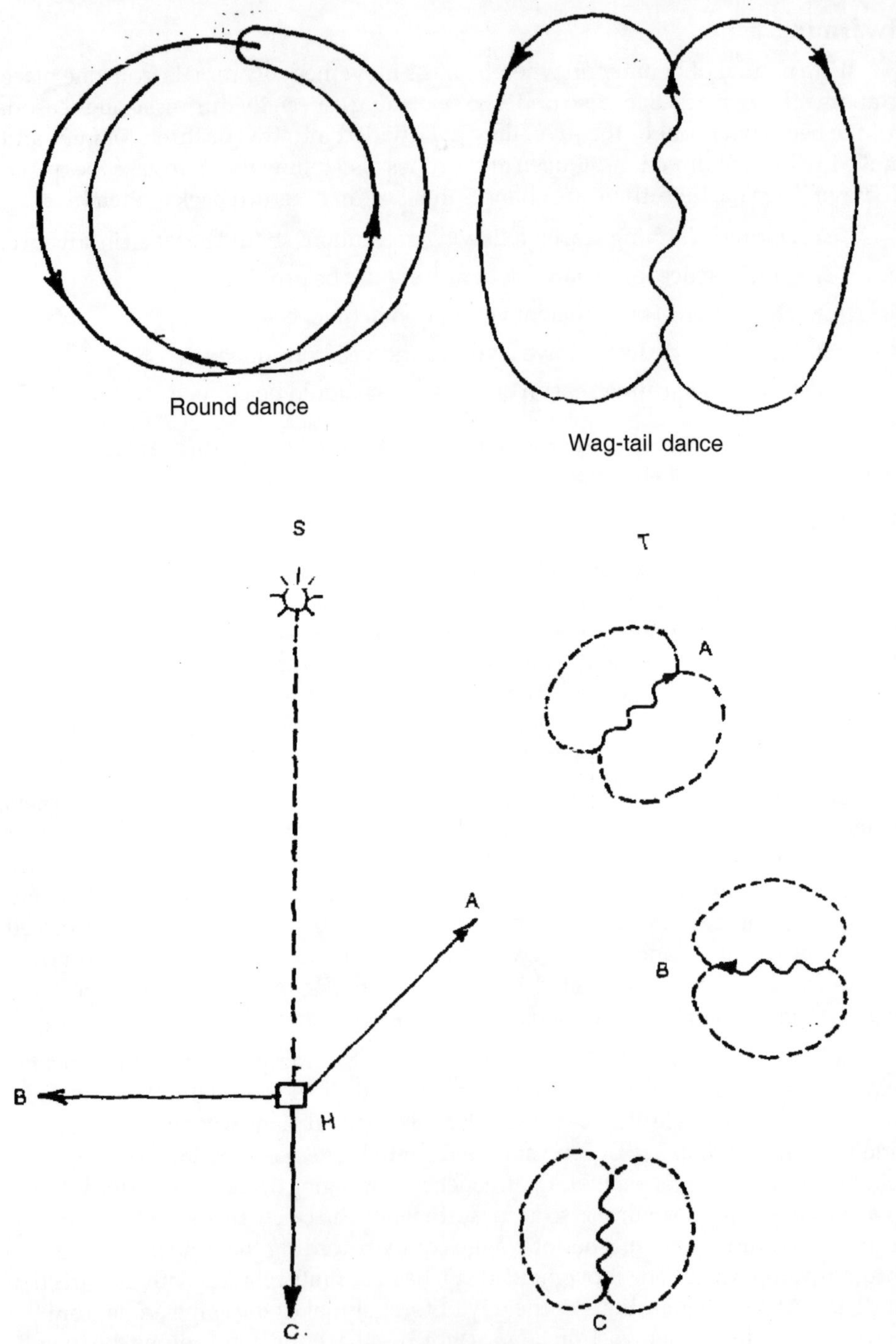

Directions of Food
S: Sun; A, B, C: Directions; H: Bee-hive

Figure 2.6: Bee Dance

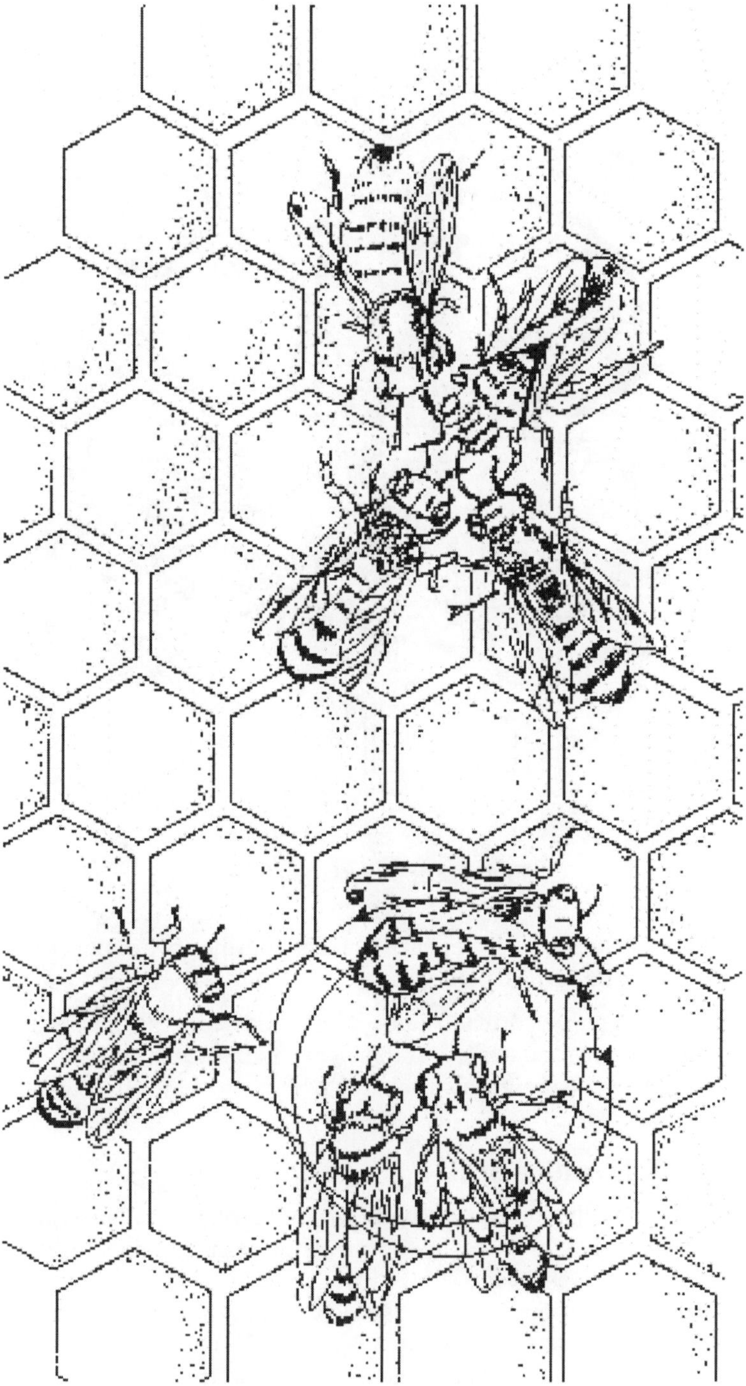

Figure 2.7: Bee Round Dance

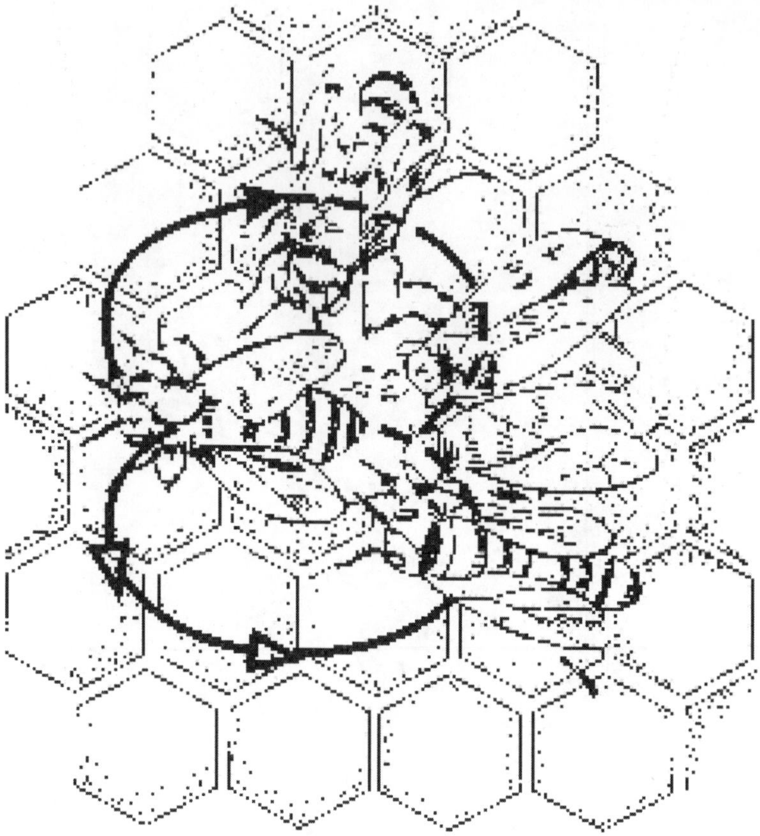

Figure 2.8: Bee Waggle Dance

If the hive, sun and food remain in a straight line, then bee dances forming a zero degree angle from the vertical plane.

When the forager performs its dance the other workers after observing it gets excited, join the dance for a while and then leave in the direction of food. The odour of pollen brought by the forager also help the workers to search that particular source of food.

Diseases and Enemies of Bee

Bees are generally thought to be resistant to diseases. This conception has developed because bees do not easily show the sign of their ill-health by their external appearance. A detailed study has shown that they do suffer from contagious diseases and very often from curious organic disorder. One of the most dreadful contagious disease found prevalent among bees are "Brood foul". The larvae suffering from this disease show uneasy movement in the cell, the colour of the body changes from pearly white to yellow, skin becomes flaccid and opaque and death occurs soon. This disease occurs perhaps due to different genera of fungi to which a common name

"Schizomycetes" is applied. It is a contagious disease and the main source of its spreading from one hive to another is man. The spores of this bacillus stick to the hands of the keeper while examining the different hives. The queen transfers these germs to their off-springs through eggs. To prevent this disease in not very acute cases, the injected hive should be sprayed by 1/150 solution of salicylic acid. Extra food is also supplied to the bees containing a mixture of salicylic acid. Some are of the opinion that the queen cells of the infected hive should be destroyed, the old queen should be replaced by a new one and the hive be painted from inside by a mixture made of two parts of methylated spirit and one part of carbolic acid crystals. This mixture destroys all bacilli and spores. Bags containing camphor placed inside the infected hive check the spreading of this disease. However, this has not been found true in all cases.

There are few other diseases which are prevalent among bees. Nosema disease and Amoeba disease are caused by protozoan parasites, *Nosema apis* and *Vahlkampfia mellifica*, respectively. Nosema affects the stomach and small intestine which becomes grayish white as compared to yellowish of a normal bee. This disease occurs throughout the world. Amoeba disease on the other hand affects the malphighian tubules of the European and American species. No successful measure of controlling either of the disease are known. "Isle of wight" or acarine disease occurs due to the blocking of the trachea of bee by a small parasitic mite, *Acarapis woodi*. A mixture of safrol, oil, nitro-benzene, methyl salicylate and petrol kill the mites of the sick bee. *Bacillu apisceptious* (Burnside) a bacteria affects the blood of bees and produces "septicemia". Methods to control this disease is not known.

There are a large number of animals who act as enemies to the bee. Wax moth destroys the comb. They are of two main types, greater wax moth-*Galleria mellonella* L. and lesser wax moth-*Achroia grisella* (Fabr). Moths lay their eggs on the combs during summer evenings. Larvae after hatching try to enter the cells of the comb and make it unable to be used any further. Healthy hive with active workers prevent their entry to a great extent. Artificially sulphur fumigation is sufficient to destroy the eggs and larvae. Calcium cyanide, carbon-disuphide, carbon tetrachloride mixture and methyl bromide are also used for fumigating the comb.

Braula caeca, a small, reddish louse acting as ectoparasite on the queen's body, may disturb the whole colony since the queen's presence is necessary for a hive. This louse sucks the blood of queen by its powerful sucking and piercing mouthparts till the queen dies. Death of the queen results in disturbance of the whole colony. The method of control is simple since the louse can be easily brushed off with the help of a soft brush from the queen's body. Besides this several species of predaceous wasps causes considerable damage to a colony. They can be controlled by destroying the wasp nests from the locality. The wasp nests can be destroyed by fumigation with calcium cyanide or by burning them with kerosene and petrol torches. To prevent the entry of wasps inside the hive the entrance guard of appropriate size should be fixed.

Wax Beetle of genus Platybolium and Bradymerus produces unhygienic condition inside hive as its grubs share the major part of debris deposited on the bottom plate. A regular cleaning of hive is advisable to minimize its effect.

Birds especially blue tit, fly-catches, chaffinch, sparrow, etc., use bee as their meal. Toads hiding beneath hives have been found to feed upon the tired workers. Snails, ants, dragon flies, praying mantids, termites etc., are other agents which annoy bees considerably.

Usefulness of Honey Bee

The most important part played by bees is the production of honey, which is a nourishment of high food value.

Honey

It is a sweet, viscous, edible fluid obtained by honey bees from nectar and pollen secreted by plants. The flight radius of the Indian honey bees is between ½ of one mile, which is much less than their Western counterparts who cover a radius of 3 to 5 miles. The collection of nectar is veryhard and strenuous work, as for collecting 500 gm of nectar a bee has to make about 10,000 flights. Out of this 500 gram about half gets evaporated. We can say that for collecting one pound of honey, bees have to make more than forty thousand trips and therefore the calculated distance covered is about double the circumstance of the earth.

When the bee sucks the nectar from the flower it passes them to its honey sac where it gets mixed with its acid secretion. These nectar and pollen are dropped in particular channels in the hive. Further processing of temperature in the hive is done by workers through flapping of their wings. The actual process of honey formation is not possible to enumerate in detail. However, it is believed that the cane sugar of the nectar is converted into dextrose and levulose inside the honey sac by the action of certain enzymes. After regurgigation it finally changes into honey which is stored in the hive for future use.

Chemical Composition of Honey

An average sample of honey is composed of water, sugar (levulose, dextrose, sucrose, dexrin), ash (minerals like calcium, iron, phosphate and manganese), about 8 components of vitamin B complex (pantothenic acid, biotin, pyridoxin, choline, ascorbic acid, thiamine, riboflavin and miacin). Besides this honey is an antiseptic and contains formic acid as the preservative. The colour, flavour and odour of honey usually depend on flowers from which nectar is gathered. One kilogram of honey contains 3200 calories and is energy rich food. It is easily digestable, hence given to the infants.

The optimum temperature for the storage of honey is 70°F, below which it looses its colour and the glucose molecule cyrstallises. The unripe honey containing more than 20 per cent water may possibly get infected by sugar tolerant yeasts resulting in the formation of alcohol, water, acetic acid and carbon dioxide. Therefore, before storing the honey it should be heated at 160°F for 30 minutes in order to remove excess water.

Bee Wax

It is a wax of high melting point (about 140°F) secreted by wax glands of worker bees. It is utilized in the construction of hive. This wax is used by human beings for several purposes like manufacturing of cosmetics, cold creams, shaving creams, polishes, candies, ointments, lipsticks, lubricants, in modeling works etc. It is also used by beekeepers in the formation of comb-foundation bases for modern bee live.

Propolis and Balms are other collections of bee from the plants. These substances are utilized in repairing and fastening the comb. Bees are good pollinators and responsible for cross pollination in several variety of leguminous plants. Thus they play a good role in agriculture and horticulture. Lastly the sting of bee which is source of annoyance to man is supposed to be the cure of few diseases. It is used in manufacturing of Ayurvedic medicines. Honey is supposed to be blood purifier, a cure against cough and cold, sore throat, ulcers of the tongue, ulcer of stomach and intestine etc. it is prescribed for heart and diabetic patients and is useful for kidney and lung disorders.

3

Sericulture

Silk has been used by human beings for various purposes since ancient times. Pure silk is one of the finest and most beautiful natural fibres of the world and is said to be "the queen of fibres". Silk clothes have a look and feeling of affluency that no other cloth can equal. Due to its great value and usefulness, there have been many attempts in various parts of the world for the large scale production of silk. One of the method was the rearing of silkworms on large scale with great care in natural and controlled conditions. Different rearing techniques are applied in different parts of the world for large scale production of silk threads of fine quality. This is known as sericulture.

History

There is no authentic information regarding the origin and use of silk. The ancient litreature gives two views. According to one view, silk industry originated for the first time in India at the foot of the Himalayas, and from there it spread to other countries of the world. Second view, which has greater acceptance, says that this industry originated in China about 3000 B.C. According to this, a Chinese Princess Siling Chi was the first to discover the art of reeling an unbroken filament from a cocoon. This art was kept a close secret for nearly 3000 years. This art later on spread to the rest of the world through several agencies like civil war refugees, war prisioners, marriage of royal families etc.

Types of Silk

Moths belonging to families Saturnidae and Bombycidae of order lepidoptera and class Insecta produces silk of commerce. There are many species of silk-moth which can produce the silk of commerce, but only few have been exploited by man for the purpose. Mainly four types of silk have been recognized which are secreted by different species of silk worms.

(*i*) Mulbery Silk

This silk is supposed to be superior in quality to the other types due to its shining and creamy white colour. It is secreted by the caterpillar of *Bombyx mori* which feeds on mulberry leaves.

(*ii*) Tasar Silk

It is secreted by caterpillars of *Antheraea mylitta, A. paphia, A. oyeli, A pernyi, A. proyeli* etc. This silk is coppery colour. They feed on the leaves of Arjun, Asan, Sal, Oak and various other secondary food plants.

(*iii*) Eri Silk

It is produced by caterpillars of *Attacus ricini* which feed on castor leaves. Its colour is also creamy white like mulberry silk, but is less shining than the latter.

(*iv*) Munga Silk

It is obtained from caterpillars of *Antheraea assama* which feeds on Som, Champa and Moyankuri.

Different types of silk and their insects along with their particular food plants are given below:

Different Types of Silkworm and their Hosts

Type of Silk	*Type of Silk Insects*	*Food Plants*
Mulbery	*Bombyx mori*	*Moras alba* (Mulbery)
Tasar	*Antheraea mylitta*	*Terminalia arjuna* (Arjun)
	Antheraea paphia	*Terminalia tomentosa* (Asan)
	Antheraea royeli	*Sorea robusta* (Sal)
	Antheraea pernyi	*Zizyphus jujuba* (Plum) etc.
Eri	*Attacus ricini*	*Ricinus communis* (castor)
Munga	*Antheraea assama*	*Tetraanthera monopetala* (Som),
		Michalia oblonga (champa),
		Listea citrata (Moyankuri)

Habit, Habitat and Life History

Out of the four different silk types mulberry and eri are manufactured from domesticated silkworms, whereas tasar and munga silkworms are wild in nature, although attempts are in progress to domesticate them too. The life-cycle of these four types of silk moths are much in common, as they ley eggs, from which caterpillars hatches. They eat, grow and produces cocoon for their protection, then pupate inside cocoon. After sometime moths emerge from the cocoon, male and female mate, lay eggs, and repeat their life-cycles.

Mulbery Sericulture

This is multivoltine. Mulberry silkworms are of domesticated type because they can be reared indoors. Large and healthy cocoons are selected during harvesting season for the next crop. These cocoons are kept in well ventilated cages. They emerge after few days. Males and females are easily distinguishable (female larger in size and abdomen broader than the males). The matured moth measures about 1½″-2″ in wing span and is pale creamy in colour. These moths are kept in pairs (one male and one female) in coupling Jars for about 24 hrs. They copulate during this period and after that the females are transferred to egg-laying boxes made up of card-board or

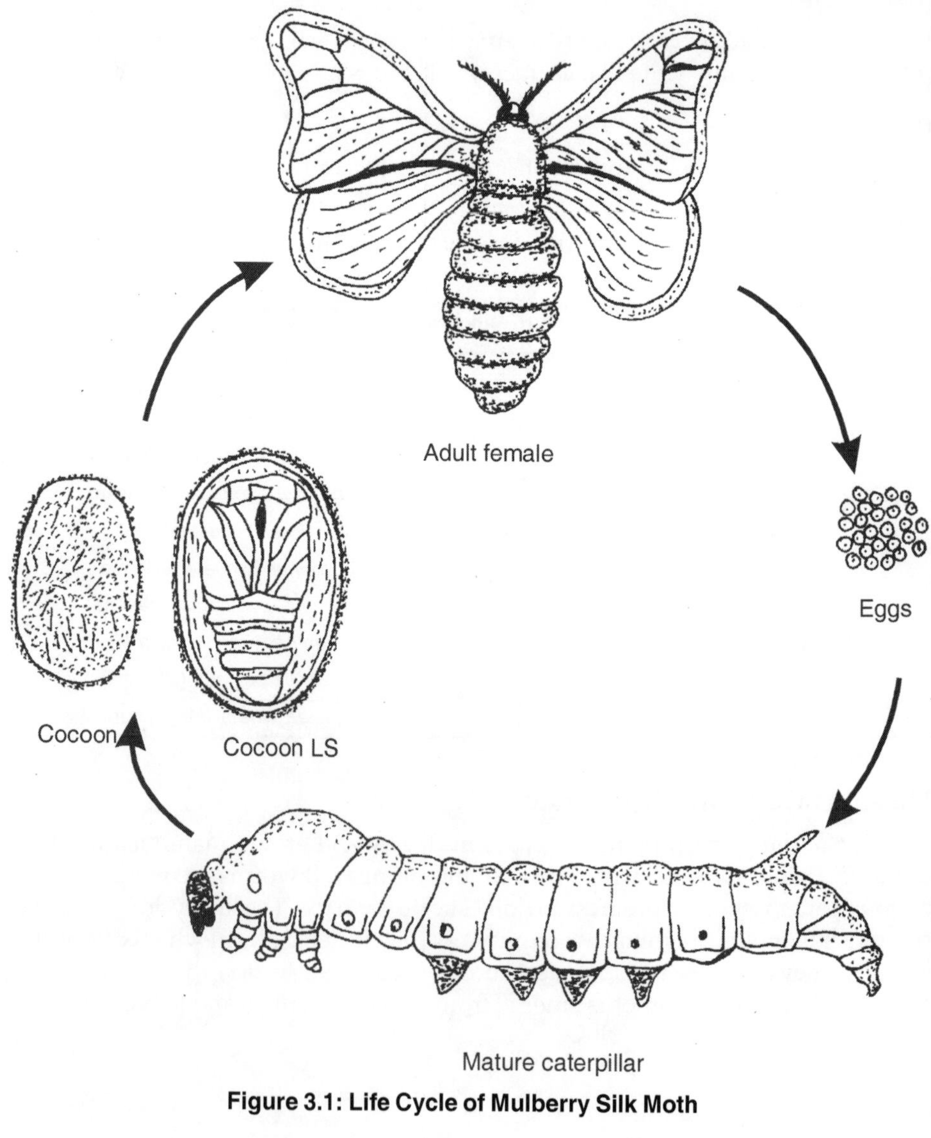

Figure 3.1: Life Cycle of Mulberry Silk Moth

earthen-ware. Each female lays about 300-400 eggs in the next 24 hrs. These eggs are very small in size, white in coour, seed like in appearance and commercially they are known as "seeds". A great care is required in selection of these seeds because the whole industry is based on these seeds. They should be healthy and free from any sort of disease. For this, tissue fluid of egg-laid moths are examined under microscope. If an evidence of any type of disease is found, the eggs laid by that particular moths are destroyed. These disease free seeds are also supplied to villagers and tribal rearers. These eggs are kept in incubators (at 75°F) in small boxes where they hatch between 7-10 days. These small larvae are known as caterpillars which are subjected to the process of rearing.

Cultivation of Food Plants

The larvae of *Bombyx mori* feed on leaves of mulberry, so for undisrupted and regular rearing a continuous supply of healthy mulberry leaves is essential. For this a systematic and regular cultivation of food plants is maintained. Although mulberry leaves have high food value, their cultivation and maintenance is a costly affair. Cost of silk production is directly proportionate to the cost of mulberry leaves. Therefore, for large scale production of silk at a cheaper rate, it is essential to reduce the cost of mulberry leaves. One of the techniques for the purpose is to raise the mulberry trees from "grafting", rather than from cuttings or seeds. They are economical because they require little attention and maintenance once they are established.

Rearing of Silkworms

Tools and materials necessary for raring silkworms:

House

Any building or thatch which is well ventilated may be used for rearing the worms, but mud-walled hatched houses are the best as they are cool in summer and warm in winter season. During summer season water may be sprinkled inside the hatch to lower the high temperature. For proper growth and development of silkworms the temperature inside the house should be maintained more or less between 70-75°F with similar percentage of humidity.

Feeding Trays

Freshly hatched worms are kept in flat trays alongwith small pieces of mulberry leaves. These trays are made up of bamboo mattings with their edges turned up, which afford a raised border made by stout stripes of bamboos. On the back of the tray two stripes are firmly fastened longitudinally.

Machan

Machans are needed to accommodate large number of trays in a limited space. Machans are easily and best made by fixing two pairs of bamboo or wooden poles in the ground and tying across bars of bamboo or wood horizontally.

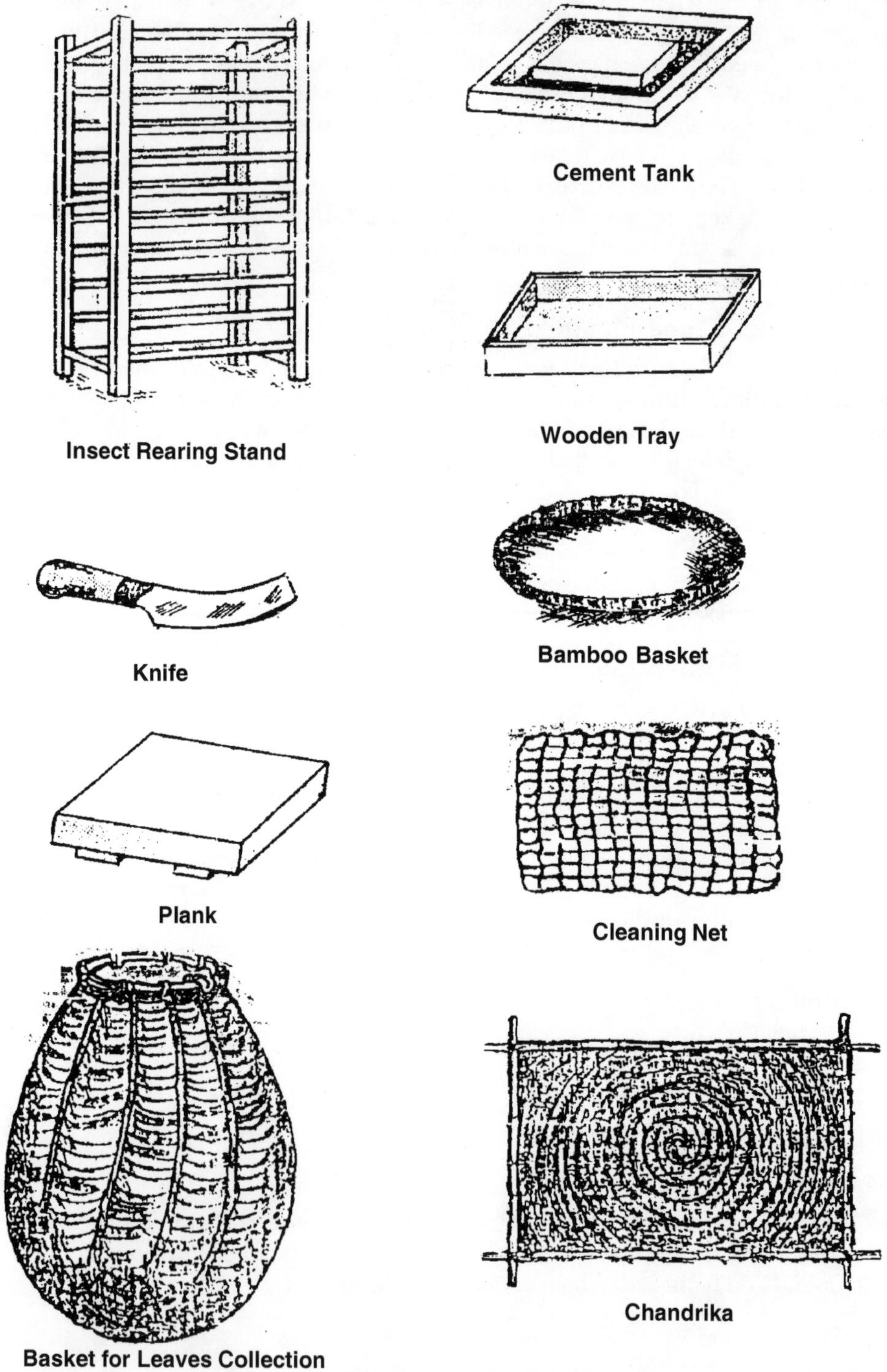

Figure 3.2a: Equipements of Silkworm Rearing

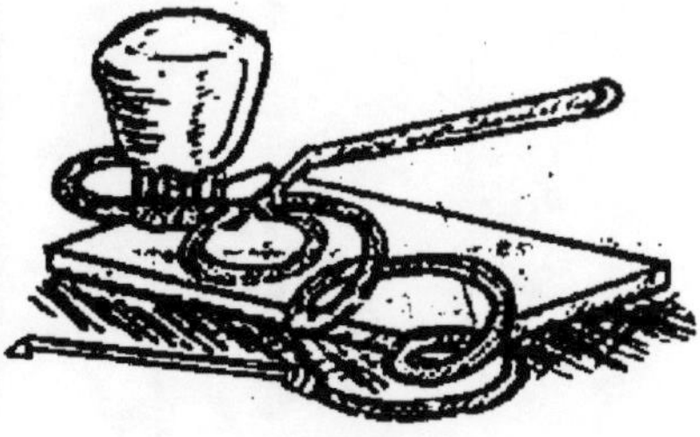

Sprayer

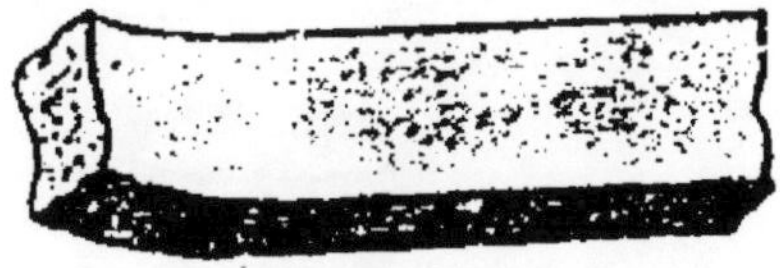

Foam Sheet

Feather

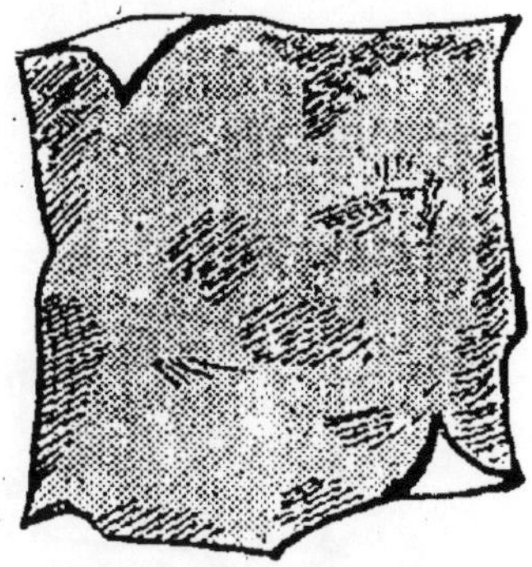

Parafine Paper

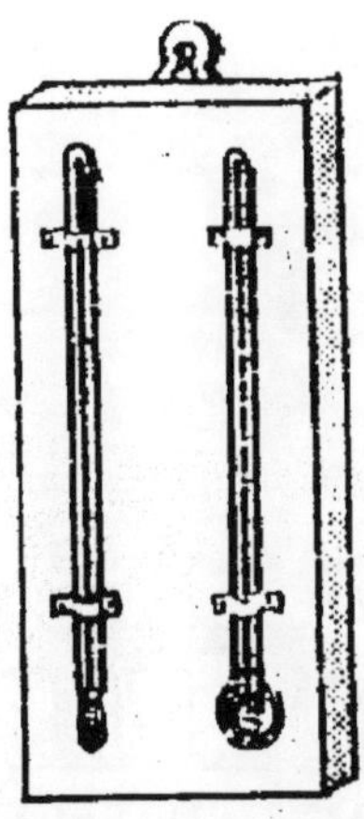

Dry-wet Thermometer

Figure 3.2b: Equipements of Silkworm Rearing

Nets

Large amount of excreta, dirty products and remains of leaves may fall on the trays from the holes of the upper trays. If worms are not protected from these by-products they may get diseased. To prevent this, trays are covered with the nets.

Spinning Trays

Before cocoon formation, mature worms are transferred to special type of trays known as spinning trays or chandrika. Here they spin the cocoon without any disturbance.

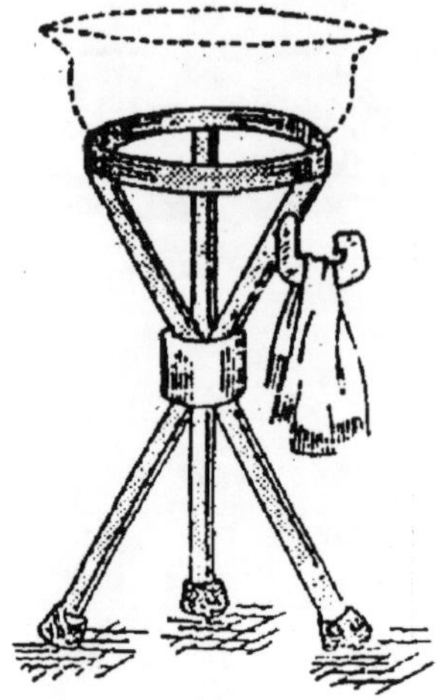

Basin Stand

Basin

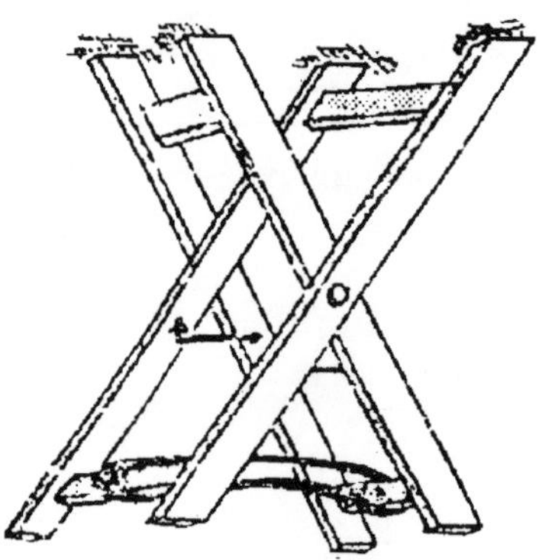

Feeding Stand

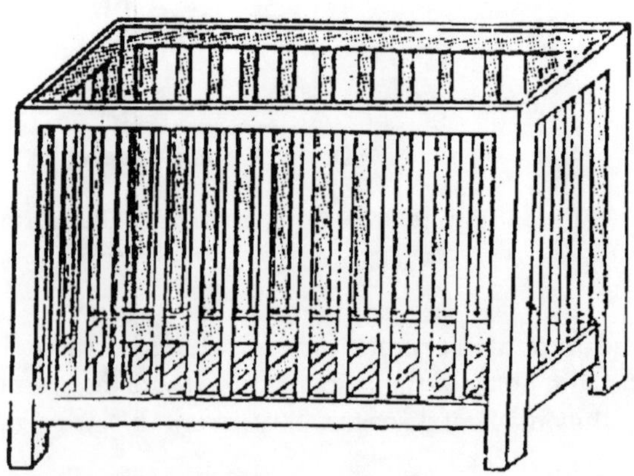

Leaf Collection Chamber

Figure 3.2c: Equipements of Silkworm Rearing

Copper Sulphate, Sulphur and Some Other Germicides

Before the start of rearing the machan, feeding trays, spinning trays, nets and everything used in rearing except the leaves are washedwith copper sulphate ($CuSO_4$) solution or other antiseptic chemicals. The remaining germs are further killed by fumigation of sulphur.

Tiny caterpillars which hatches from the egg measures 5-7 mm in length. They are transferred to feeding trays already supplied with chopped tender leaves of mulberry. These caterpillars move on the leaves in a characteristic looping manner. Their body is rough, wrinkled grayish in colour. They are made up of 12 segments which is distinct into three parts *i.e.*, head, thorax and abdomen. The head bears mandibulate mouth-parts with which they feed upon the leaves. The thorax is 3 segmented and all the segments bear a pair of true jointed legs. The abdomen which has 10 segments is provided with five pairs of unjointed, stumpy *prolegs* or *pseudolegs*. (One pair each in segment 3rd, 4th, 5th, 6th and 10th) a short dorsal and horn (on the 8th segment) and a series of *spiracles* on lateral sides. These larvae feed voraciously upon the mulberry leaves and grow very quickly. They stop feeding, become inactive after four to five days, and then I^{st} moulting takes place. The 2nd stage larvae resembles the I^{st} stage larvae except that they are slightly bigger in size. They also eat voraciously for 7 days, then 2nd moulting takes place and 3rd stage larvae are formed. The larvae repeats this process for 4 times. The maturity is achieved in about 45 days since the time of hatching and the matured caterpillar now measures 7-10 cms in length. By this time the formation of a pair of salivary glands is completed. Since these *salivary glands* secrete silk they are also called as *silk-gland*.

When the matured Caterpillars stop feeding they are transferred to *spinning trays*. They excrete their last excreta and begin to secrete the sticky secretions from the silk-gland through a very narrow pore situated on the hypopharynx. The secretion is continuous and after coming in contact with the air sticky secretion is converted into a fine, long and solid thread of silk. The thread becomes wrapped around the body of larva forming a *pupal case or cocoon*. This process continues for 3-4 days, at the end of which the caterpillar is enclosed within a thick, somewhat hard, oval, whitish or yellowish cocoon. Within 15 days the caterpillar is transformed into a brownish *pupa or chrysalis*. Active metamorphic changes takes place during pupation in which abdominal prolegs disappear, while the thorax develops two pairs of wings.

The pupa is finally metamorphosed into young adult moth in about 12-15 days. This young moth or imago secretes an alkaline fluid to soften one end of the cocoon and then escapes by forcing its way out of the softened silk. Soon after the emergence, the silk moths mate, lay eggs and die. Just after the formation of cocoons healthy cocoons are selected and kept in cages for the next crop.

The following precautions should be taken during the rearing of mulberry silkworms:

1. The worms should never be kept overcrowed in a tray.
2. Dried or dusty wet leaves should never be fed to the worms.

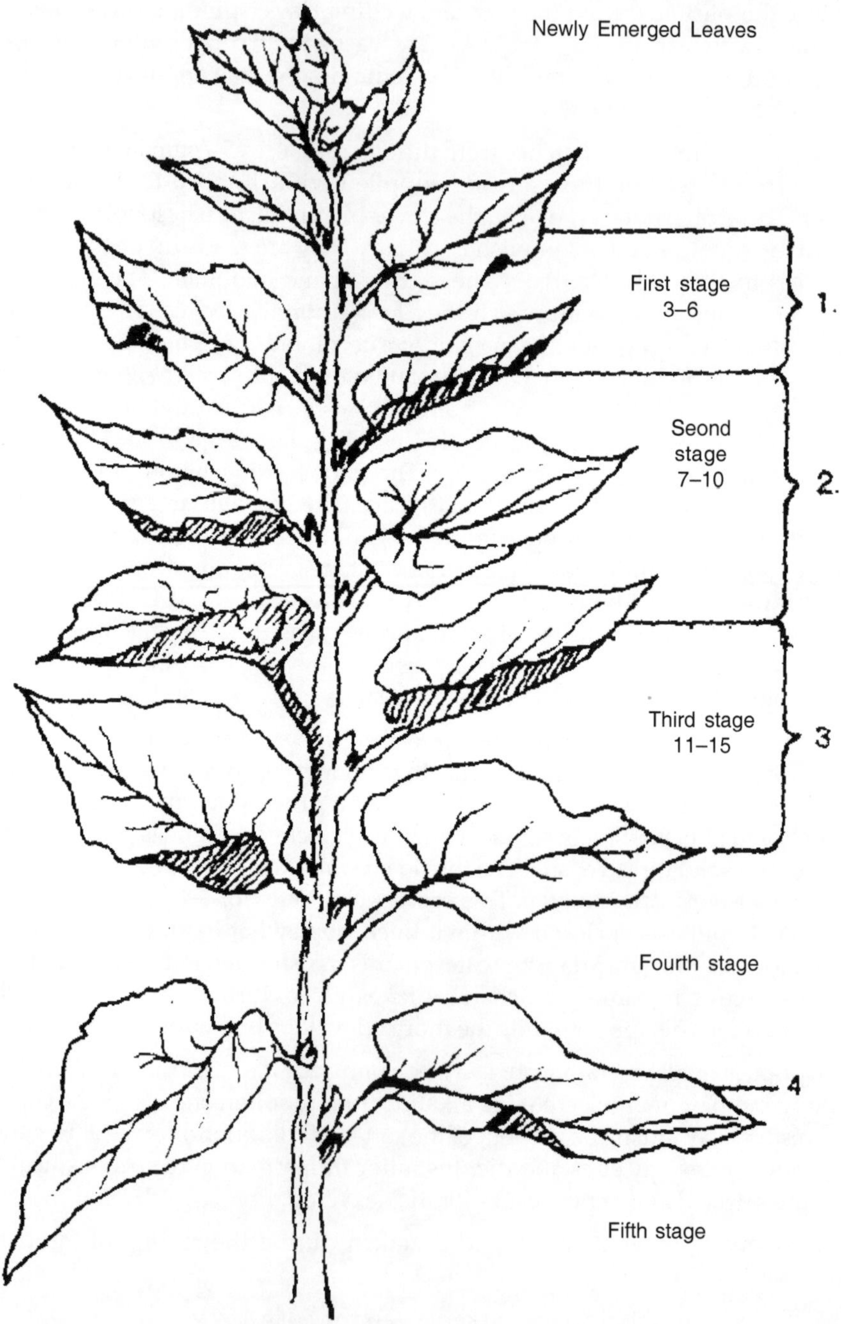

Figure 3.3: Selection of Leaf for Rearing, Various Stages of Mulberry Silkworm

3. Proper ventilation is a must but wind should not be allowed directly over the worms.
4. There should be equal distribution of leaves among the worms.
5. Worms which are under process of moulting should not be distured otherwise they may die or moulting may be delayed.
6. There should be no dust at the floor of the house. For this it should be well plastered with cowdung or mud at regular intervals.
7. Smoking should be strictly prohibited in the rearing room.
8. Worms should not be handled with dirty hands otherwise they may get diseased. They should be handled only after washing the hands thoroughly with antiseptic solution and drying the hands.
9. One should enter the rearing house only after putting off the shoes, etc.
10. During hot or summers, drinking water may be sprinkled over the feeding trays.

Reeling of Raw Silk From Cocoon

Before reeling the thread the cocoons are dipped in a container of hot water for more than 10 minutes. During this period they are continuously stirred with a rod. Due to this, their outer portion is loosened and removed in the form of long tapes and the end of the continuous filament is found. The filaments of several cocoons are picked up and passed through the 'glass eye' on the reel. The thread thus reeled forms the 'raw silk' of commerce. About 1 kg. of raw silk is obtained from nearly 55,000 cocoons.

Tasar Sericulture

Cultivation of Food Plants

Since they are wild in nature, cultivation of food plants is not at all necessary. The worms are mounted on the food plants in the nearby forests. But still for convenience plantation of food plants can be done. Their primary food plants are Asan, Arjun, Sal, Oak etc. and there are large number of secondary food plants.

For the cultivation of the food plants first of all particular piece of land selected for the purpose is prepared (ploughing, leveling, manuring etc.) and then the saplings are implanted there after sufficient rain fall. The distance between two saplings should be 20-25 ft. Watering, manuring, and ploughing of the soil around the saplings are done at regular intervals according to the need. They are protected from cattles and other animals. Proper care is taken till they have attained a considerable height. Bushes are pruned 3-5 weeks before the start of the rearing. It is not advisable to rear the worms on a plant every year because in that case sufficient foliages will not be available for the developing larvae. To overcome this problem, the land in which rearing is to be done is divided into two plots. In a particular plot rearing should be done every alternate year. For speedy and healthy growth of the offshoots, it is necessary to give proper care and attention to host plants which includes ploughing, manuring, watering and pruning at regular intervals.

Rearing of Silkworms

They are bivoltine *i.e.,* two crops in a year, one from August-October and other from October-December. It is from August to December that the tasar insects are active and for the rest of the year they are inactive *i.e.,* under dipause. The active and inactive phases of life of tasar insects is controlled by environmental and hormonal factors. The moulting hormone *ecdysone* plays a vital role. Ecdysone is secreted by prothoracic gland. When their secretion stops then moulting stops, worms become inactive, and when secretion starts worms become active again.

Healthy cocoons are selected during the harvesting season for the next rearing. These cocoons are kept in well ventilated cages. In favourable season emergence of male and female moths take place. The size of the tasar silk-moths is larger than the

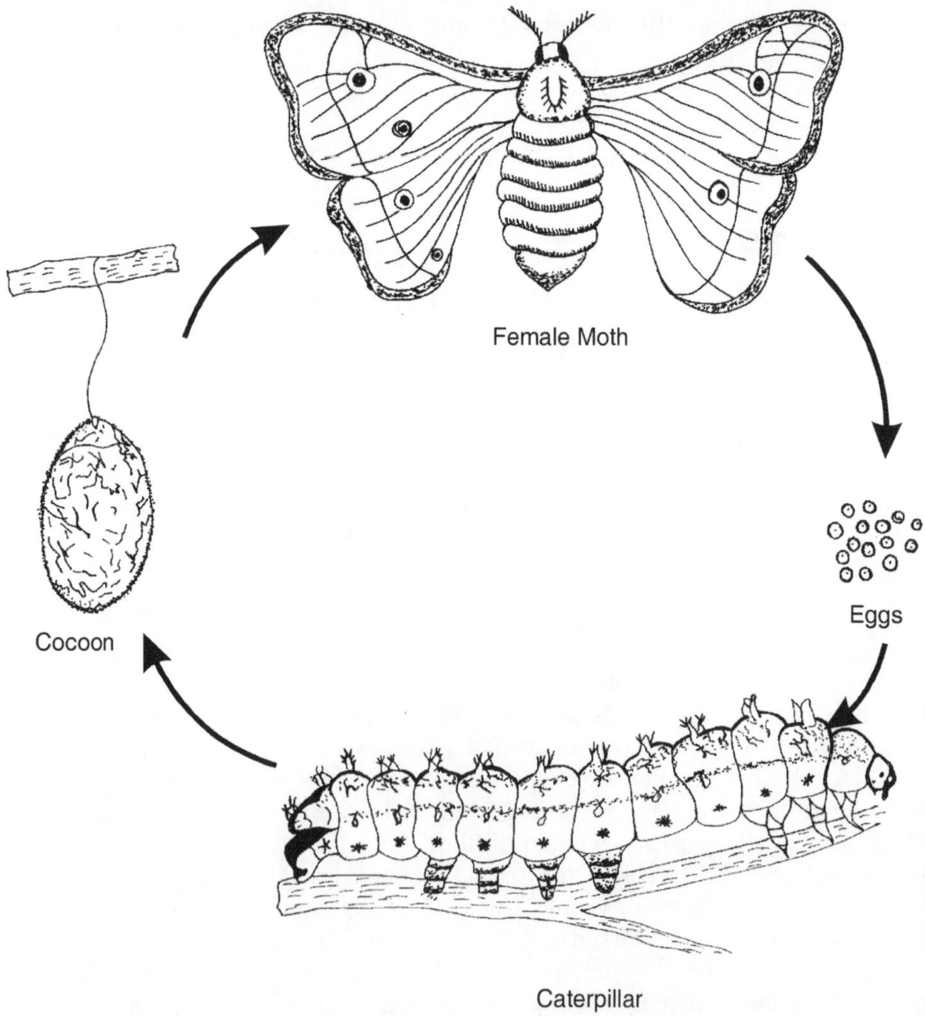

Figure 3.4: Life Cycle of Tasar Silkworm

other silk-moths. The female which is yellowish or deep brown in colour is larger than the brick-red coloured male. Wings generally have an eye-spot. For successful mating monias made up of palm leaves are used. In each monia a pair of moth is kept. Within 24 hrs. they are expected to complete their mating process. After completion of this period, monias are opened and males are allowed to fly away. Females are transferred to earthenware pots or card-board boxes in which they deposit their eggs in next 24 hrs. These egg laid females moths are subjected to pathological test. If they suffer from any disease, the eggs laid by the particular moth is destroyed. Only eggs of disease-free moths are kept for hatching. The eggs are washed in 5 per cent formaline solution and then kept in incubators in small hatching boxes. In about 7-10 days the egg hatch.

The freshly emerged larvae are yellowish in colour, covered with bristles and measures less than ½″ in length. They are mounted on the bushes of the host-plants already pruned for the purpose. With the help of their mandibulate type of mouth-parts they feed on the tender leaves of host-plants. They grow in size and after 3-4 days they become inactive and Ist moulting takes place. Structural details of caterpillars of tasar-worms are similar to that of the mulberry-worms with slight variations. With successive moults the size increases and colour changes. There occurs four moultings in the life-cycle of a tasarworm and thus five larval stages exists. The fifth larval stage is of the longest duration (15-20 days) and it measures approximately 4″-5″ in length and 50 gms in weight. It takes about 40-50 days for a freshly hatched larvae to attain the size of fullgrown and healthy caterpillar which is capable of spinning a cocoon.

After passing out the last excreta the larva takes rest for a while, then it becomes active in search of a suitable place for spinning the cocoon. After selecting a suitable position for the formation of ring, which is generally above a node, the caterpillar crawls down to form a *hammock* by tying a few leaves with silk-threads. The *hammock* is generally in the shape of a cone or a cup with an opening on the top. After hammock formation larvae comes out from the hammock to form a *ring* around the twig. First of all, peeling of the bark in a circular pattern takes place with the help of a pair of powerful mandibles. Around this scare the silk is thrown in a semicircular manner and within few minutes a strong ring of silken thread is formed. Ring formation is followed by the formation of the *peduncle.* Soon after ring and peduncle formation larvae enters the hammock and starts spinning the cocoon. The spinning of cocoon is completed after 4 to 6 days of its commencement and the larvae inside pupates after another 4 to 6 days.

Cocoon is a hard protective covering secreted by the silk-glands of the larva. Tasar cocoons have three parts; rings, peduncle and main body of the cocoon. These cocoons are very hard. During rearing season a continuous watch is maintained to prevent the tasar worms to be eaten up or destroyed by insectivorous birds, bats, rats, squirrels, lizards, predators, parasites etc.

Reeling of Raw Silk From the Cocoon

The mechanism is slightly different than the mulberry ones due to hard nature of the cocoons, which is due to presence of a gummy substance. Reeling of the silk filament cannot be done unless this dried gummy substance is softened. For this special cooking technique is employed. The cocoons are dipped in 0.5 per cent sodium carbonate (Na_2Co_3) solution for 18 hrs. Now the cocoons are subjected to stema-cooking under 15 lb./inch^{-2} pressure for 2½ hrs. For better tensile strength of the filament these cocoons after 24 hrs. are treated with 0.5 per cent formaldehyde solution for 15 minutes. Excess of water from the cocoons is removed by squeezing. The reeling of the silk filaments from these can be done with improved reeling machines. Four cocoons are used constantly as long as the reeling continued for each spindle.

Sericulture Eri

This type of silk is obtained from the worms of *Attacus ricini* which feeds on castor leaves. After mulberry silkworms it is the another type which can be artificially

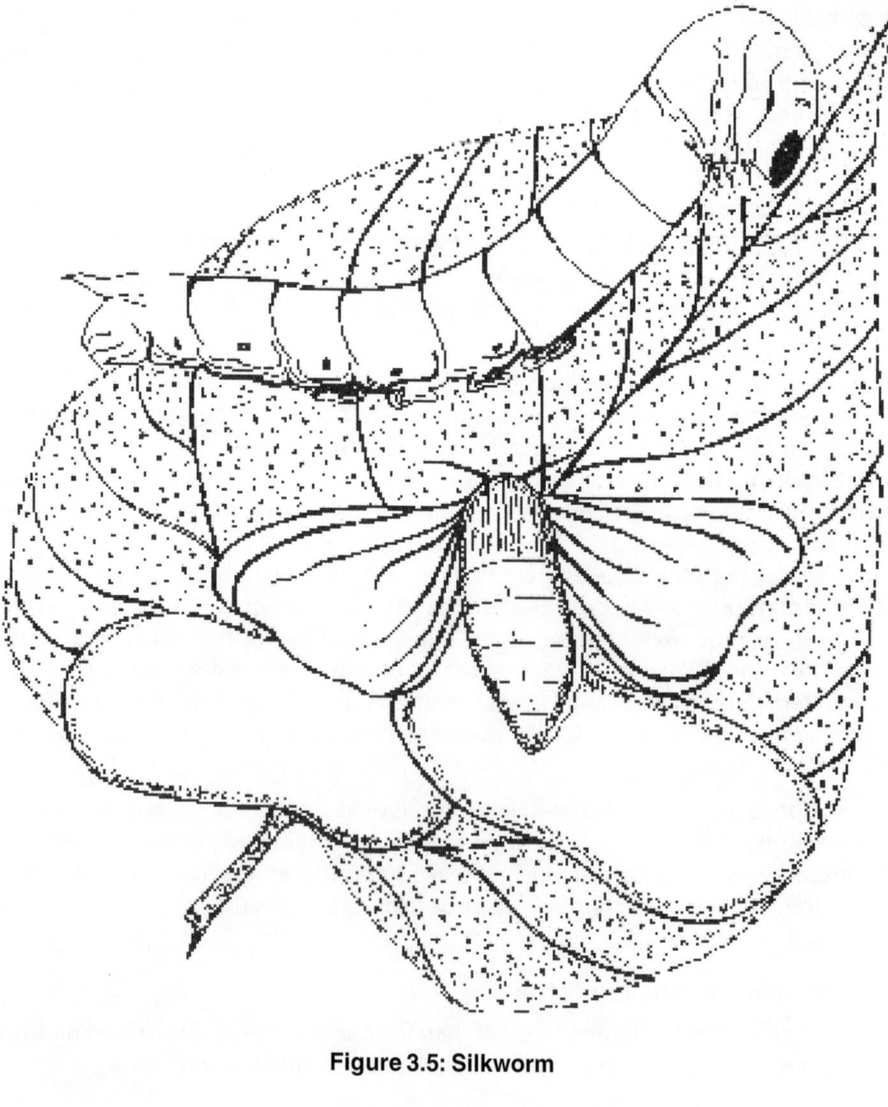

Figure 3.5: Silkworm

domesticated. But they differ from mulberry in two respects. Firstly, the silk filament is not continuous and secondly, the cocoon is made up of two layers.

Cultivation of Food Plants

Cultivation of castor plants is done like the other crops and this can be done on a larger scale. Plants are not allowed to bear fruits and seeds, with the help of regular pruning.

Rearing of Silkworms

Not only is the life-history of *Attacus* more or less similar to that of *Bombyx* but also the rearing technique, tools and materials are the same. On an average 75-85 eggs laid by a single female. They hatchout in about 8-12 days. Young caterpillar are of greenish-yellow in colour with black spots. These caterpillars are transferred to feeding trays and are supplied with small pieces of fresh castor leaves. They feed, grow and moult. The larvae undergo four moults and thus there are five larval stages. After the fourth moult, the caterpillar eats voraciously and becomes fully grown. Just before spinning cocoons they stop feeding and excrete the whole excreta from the alimentary canal. Now they are ready to spin cocoons. At this stage they are transferred to spinning baskets containing dry leaves, straws etc. Worms spin cocoons around themselves and then after sometime they pupate inside the cocoons. During rearing the same precautions are taken as described in the rearing of mulberry silkworms.

Spinning of Raw Silk from the Cocoon

Extraction of silk thread from the cocoon is different here than the other silk types. This is due to the peculiar structure of the cocoon in two respects. Firstly, the filaments of the silk are not continuous but has numerous breakages. Therefore, in this case, moths are allowed to emerge from the cocoons in the natural way. Secondly, the cocoons are double layered-inner contains cast off skin of pupa and excretory remains of the larvae and thick and hard outer layer contains the silk threads. Due to the discontinuous nature of the silk filaments, it is not reeled but spun from the outer layer of the cocoon. For removing the dirty inner portion of the cocoons, a reversing machine is used which turns the cocoon inside out. These reversed cocoons are washed several times alternately with cold water, hot water containing caustic soda, and cold water. After this, spinning is done with the help of various machines like, Pusa continuous machine, Takli or Charkha.

Diseases and Enemies of the Silk-worms

This profitable industry is often threatened not only by various diseases resulted from the viral, fungal, bacterial and protozoan infections but also by insect predators, birds and other higher animals. Diseases are of the following types:

(*i*) Pebrine

It is the most serious disease of the silkworms and if not checked will wipe out the entire industry. It is caused by an internal protozoan parasite (*Nosema bombycis*) which is contagious as well as hereditary. To cope with this serious disease body tissue fluid of the female moths should be examined under microscope. If pebrine spores are found, the eggs laid by those females should be destroyed. If the infections is of a mild nature then treatment of eggs with warm water (47°C) will serve the purpose.

(*ii*) Flacherie

It is a bacterial infection which generally occurs due to unhealthy and dirty conditions of silk house. Prevention of this disease can be done by improving the conditions of the silk houses like better ventilation, keeping it neat and clean and provision for sufficient light.

(*iii*) Grassarie

It is a sort of viral infection also caused by unhealthy conditions. General cleanliness and separation of infected worms will serve the purpose. Good food enables the worms to resist these diseases to a greater extent.

(*iv*) Muscardine

It is due to fungal infection which proves fatal. It can be controlled by using disinfectants.

Various types of silkworms diseases, their symptoms and methods of control are given in the following chart:

Name of Disease	*Nature of Disease*	*Symptoms of Disease*	*Controlling Methods*
Pebrine	Protozoan (Nosema) transmitted through the eggs	Death of under nourished larvae, Production of lesser and defective silk, death of larvae	Treatment of eggs with water (47°C) rejection of infected eggs
Flacherie	Bacterial infection	Softening of skin and physical disability	Careful rearing in healthy conditions
Grassarie	Viral infection	Moulting is affected, skin becomes yellow and blood becomes milky, putrification of internal organs and death of larvae	General cleanliness, separation of affected worms and immediate disposal of dead worms
Muscardine	Fungal infection	Worms body become hard	Maintenance of stock which are resistant to the disease and by using disinfectants

Predators

1. *Tricholiga bombycis*: This parasite fly causes great loss and is a serious menace for silk industry. They lay their eggs near or inside the body of the caterpillar. Larvae upon hatching feed upon the tissue of the caterpillar which ultimately dies.

2. *Canthecona fincellata, and tachnid fly*: They are long and stout rostrum, with the help of which, they penetrate the body wall of silk worms. A good amount of haemolymph oozes out and worms ultimately die.
3. Ants, hornets, crows, kites, bats, rates, squirrels and lizards feed upon silk worms thereby causing great loss to the silk industry.

Production of Silk

Silk is the result of secretion of silk glands. They are a pair of long tubular and coiled glands lying one on each side of alimentary canal of the caterpillar. *Fibroin,* a sort of fibrous protein is secreted by each gland which at first in the fluid condition. These glands are connected with a very narrow tube like structure known as spinneret which is a part of the hypophyarnx. The liquid secretions of two glands passes through the spinneret which transform them into a single thread. *Sericin* which causes the two fibres of fibroin to unite is secreted by a pair of accessory gland situated at the anterior region of silkgland. Two streams of fibroin along with sericin are expelled through the spinneret due to contraction and expansion of the body of caterpillar. This sticky secretion after coming in contact with the air is converted into a fine, long and solid thread of silk.

Properties of Silk

Silk threads are very fine, soft and light in weight. They are very thin but strong having high elastic property. When a cross section of the silk thread is observed under microscope, it is roughly in the figure of 8. Main inner portion (70-80 per cent) is made up of fibroin (true fibre) which is surrounded by a thin covering of *sericin* (gum covering). There is also little quantity of waxy and colouring material. Both fibroin and *sericin* are proteinous in nature. Fibroin is insoluble in water and is made up of glycine, alanine and tyrosine. *Sericin* is easily soluble in water and is composed of *sericin,* alanine and leucine.

Use of Silk

Bulk of silk fibres produced is utilized in preparing silk clothes. Uses of pure silk is decreasing gradually due to its high cost value and costly maintenance. Production of synthetic fibres has posed a serious threat to the silk industry. Clothes in which silk fibres are combined with other natural and synthetic fibres are in great demand not only in India but also in foreign countries. Seeing this demand many textile industries are manufacturing clothes like teri-silk, contsilk, etc. Besides silk being used as garments it is also used in other industries and for military purposes. It is used in the manufacture of fishing fibres, parachutes, catridge bags, insulation coils for telephones and wireless receivers, tyres of racing cars, filter clothes for flour mills, and in medical dressing and suture materials.

4
Lac Culture

Lac is a natural resin of animal origin. It is secreted by an insect, known as lac-insect. In order to obtain lac, these insects are cultured and the technique is called lac-culture. It involves proper care of host plants, regular pruning of host plants, propagation, collection and processing of lac.

History

Lac has been used in India from time immemorial for several purposes. From the epic of Mahabharat it has been recorded that Kauravas built a palace of lac for the destruction of Pandavas. We come across references of lac in the Atharvaveda and Mahabharata, so it can be presumed that ancient Hindus were quite familiar with lac and its uses.

Scientific study of lac started much later. In 1709 Father Tachard discovered the insect that produced lac. First of all Kerr (1782) gave the name *Cocus lacca* which was also agreed by Ratzeburi (1833) and Carter (1861). Later Green (1922) and Chatterjee (1915) called the lac-insect as *Tachardia lacca* (kerr), Finally, the name was given as *Laccifer lacca*.

Systematic Position

A number of species of lac insects are known, of this *Lacifer lacca* is by far the most important and produces the bulk of the lac for commerce. It belongs to:-

Phylum	–	Arthropoda
Class	–	Insecta
Order	–	Hemiptera

Super-family	–	Coccidae
Family	–	Lacciferidae
Genus	–	*Laccifer*
Species	–	*Lacca*

Food Plants

The insects live as a parasite, feeding on the sap of certain trees and shrubs. The important trees on which the lac insects breed and thrives are:

- ☆ Kusum (*Schleichera trijuga*)
- ☆ Palas (*Butea frondosa*)
- ☆ Ber (*Zizyphus jujuba*)
- ☆ Babul (*Acacia arabica*)
- ☆ Khair (*Acacia catcchu*)
- ☆ Arhar (*Cajanus indicus*)

Before coming to the actual mechanism of lac secretion and its processing, it is advisable for a lac-culturist to have detailed knowledge of lac insect and its life cycle. The adult lac insect shows a marked phenomenon of sexual dimorphism. The male and female insect vary in shape, size and also in presence or absence of certain body parts.

Structure of Male Lac-insect

It is larger in size and red in colour. The body is typically divided into head, thorax and abdomen. The head bears a pair of antennae and a pair of eyes. Mouth parts are absent so a male adult insect is unable to feed. Thorax bears three pairs of legs. Wings may or may not be found. Abdomen is the largest part of the body bearing a pair of caudal setae and sheath containing pennies at the posterior end.

Structure of Female Lac-insect

It is smaller in size. Head bears a pair of antennae and a single proboscis. Eyes are absent. Thorax is devoid of wings and legs. The loss of eyes, wings, and legs are due to the fact that the female larvae after settling down once, never move again and thus these parts become useless and ultimately atrophy. Abdomen bears a pair of caudal setae. It is female lac insect which secrets the bulk of lac for commerce.

Fertilization

After attaining the maturity, males emerge out from their cells and walk over the lac incrustations. The male enters the female cell through anal tubular opening and inside female cell it fertilizes the female. After copulation, the male dies. One male is capable of fertilizing several females. Females develops very rapidly after fertilization. They suck more sap from plants and exude more resin and wax.

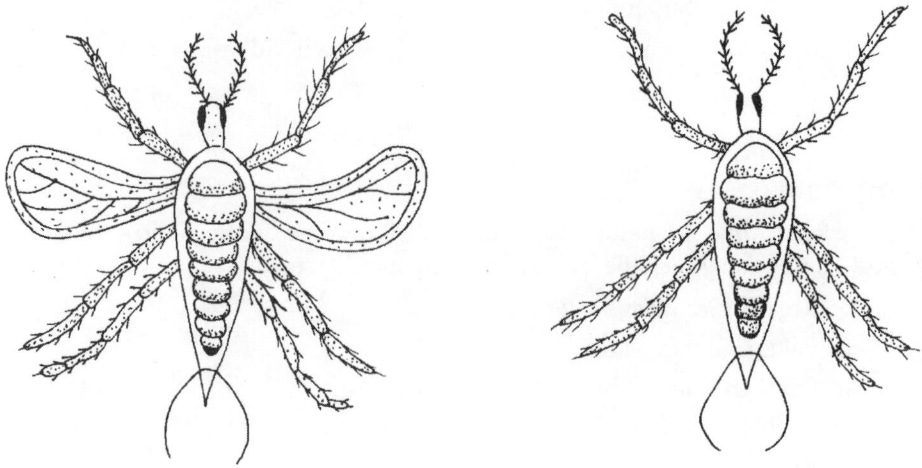

(A) Male with wings (b) Male without wings

Figure 4.1: Structure of Male Lac-Insect

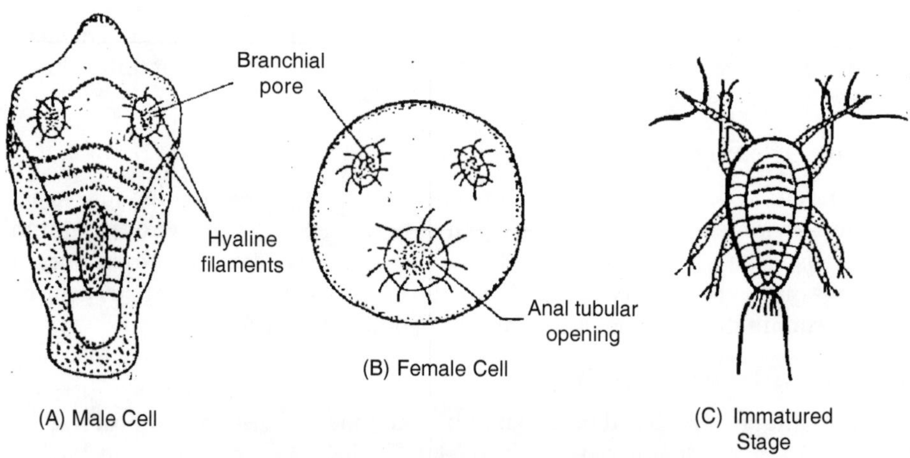

Figure 4.2: Different Forms of Lac Cells

Life Cycle

The females after fertilization are capable of producing eggs. But it has been noticed in case of lac insects that the post fertilization development start when the eggs are still inside the ovary. These developing eggs are oviposited into the incubating chambers (formed inside the female cell by the body contraction of females). A female is capable of producing about one thousand eggs (average 200-500). Inside incubating chamber, the eggs hatch into larvae.

The larvae are minute, boat shaped, red coloured and measure little over half milimetre in length. Larva consists of head, thorax and abdomen. Head bears a pair of antennae, a pair of simple eyes and a single proboscis. All three thoracic segments are provided with a pair of walking legs. Thorax also bears two pairs of spiracles for respiration. Abdomen is provided with a pair of caudal setae.

These larvae begin to wander in search of suitable center to fix themselves. This mass movement of larvae from female cell to the new off-shoots of host plant, is termed as "swarming". The emergence of larvae from female cell occurs through anal tubular opening of the cell and this emergence may continue for three weeks. The larvae of lac are very sluggish and feed continuously when once they get fixed with the twig. In the meantime the larvae start secreting resinous substance around their body through certain glands present in the body. After some-time the larvae gets fully covered by the lac encasement, also known as lac cell. Once they are fully covered, they moult and begin to feed actively. The cell produced by male and female differ in

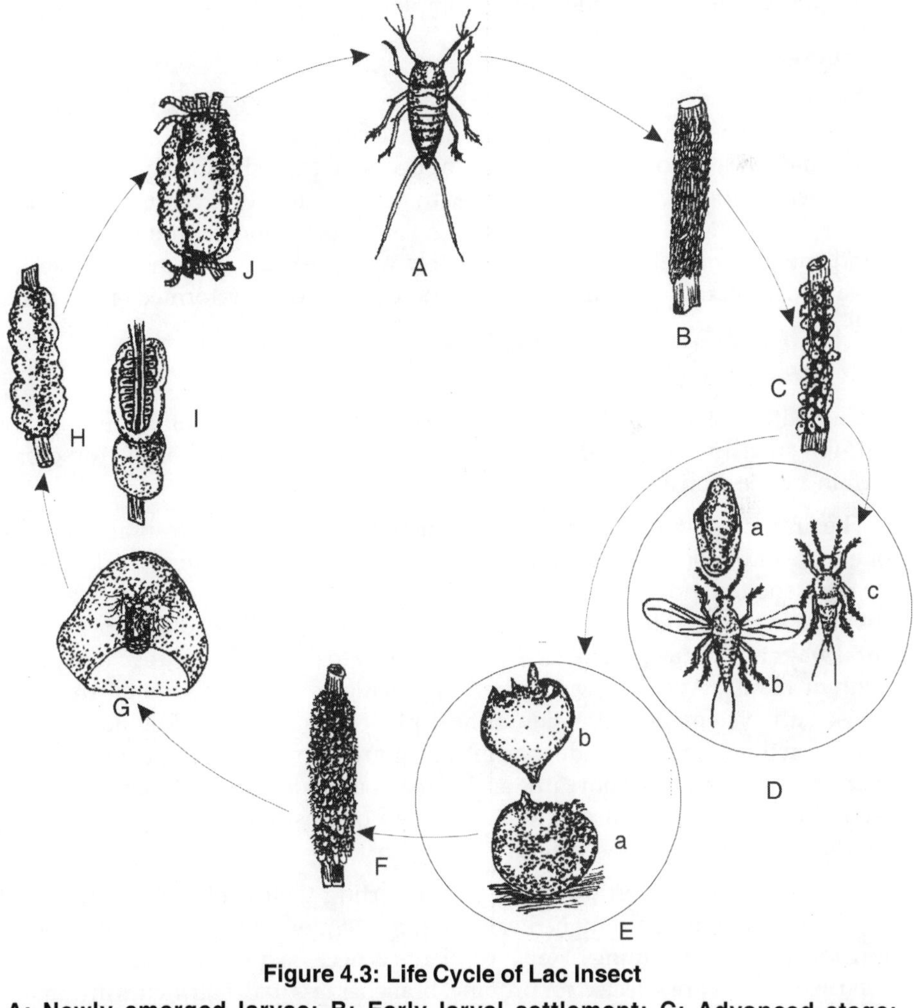

Figure 4.3: Life Cycle of Lac Insect

A: Newly emerged larvae; B: Early larval settlement; C: Advanced stage; D: The male; E: The female; F: Lac encrustation after male emergence; G: Female cell; H, I: Stick bearing encrustation of newly mature female lac insect; J: Broodlac stick tied together for inoculation purpose.

shape, and can be easily distinguished sometimes later. Male cells are elongated and cigar shaped. There is a pair of branchial pores in the anterior side and a single large circular opening covered by the flap in the posterior side. It is through the posterior circular opening that the matured male lac insect emerges out of its cell. Female cell is oval, having a pair of small branchial pores in anterior side and a single round anal tubular opening in posterior side. Through the anal tubular opening are protruding waxy white filaments, secreted by the glands in the insects body, which is an indication that the insect inside the cells is alive and is in healthy condition. These filaments, also prevent the blocking of the pore during excess secretion of lac.

Larvae moult in their respective cells. It is the second stage larva which undergoes psedupupation for a brief time, whereby it changes into adult stage. Now the male emerges out from its cell, moves on lac incrustation and enters the female cell for fertilization. In this way the life cycle is completed.

Lac Secretion

Lac is a resinous substance secreted by certain glands present in the abdomen of the lac insects. The secretion of lac begins immediately after the larval settlement on the new and tender shoots. This secretion appears first as a shining layer which soon gets hardened after coming in contact with air. This makes a coating around the insect and the twig on which it is residing. As the secretion continues the coating around one insect meet and fuses completely with the coating of anotehr insect. In this way a continuous or semi-continuous incrustation of lac is formed on the tender shoots.

Cultivation of Lac

Cultivation of lac involves proper care of host plants, regular pruning of host plant, infection or inoculation, crop-reaping, control of insect pests, forecast of swarming, collection and processing of lac.

The first and perhaps the most important prerequisite for cultivation of lac is the proper care of the host plants. It is the host plants on which lac insects depend for their food, shelter and for completion of their life cycle. There are two ways for the cultivation of host plants. One is that plants should be allowed to grow in their natural way and the function of lac-culturist is only to protect and care for the proper growth of plants. Another way is that a particular piece of land is taken for the purpose and systematic plantation of host plant is made there. Regular watch is necessary in this case by providing artificial manures, irrigation facilities, ploughing and protecting the plants from cattle and human beings for which the land should be fenced. The larvae of lac insects are inoculated on host plants only after the host plants have reached a proper height.

The lac larvae feed on the cell sap by inserting their proboscis in the tender twigs. The proboscis can only be inserted in the tender young off-shoots. For this before inoculation, prunning of lac host plants is necessary. The branches less than an inch in diameter are selected for pruning. Branches less than half inches in diametre should be cut from the very base of their origin. But the branches more than half inch diametre should be cut at a distance of 1½ inch from the base.

Inoculation

The method by which the lac insects are introduced to the new lac host plant is known as inoculation. This may be of two types, namely "Natural infection" and Artificial infection".

When infection from one plant to other occurs by natural movements of insect, it is called natural infection. This may be due to overcrowding of insect population and nonavailability of tender shoots on a particular tree.

Artificial infection takes place through the agencies other than those of nature. Prior to about two weeks of hatching, lac bearing sticks are cut to the size of six inches. They are called "brood lac". Brood lac are then kept for about two weeks in some cool place. When the larvae start emerging from these brood lac, they are supposed to be ready for inoculation. Strings could be used for tieding the brood lac with the host plant, may be of different types. In longitudinal infection the brood lac is tied in close contact with host branches. In lateral infection the brood lac is tied across the gaps between two branches. In interfaced method, brood lac are tied among the branches of several new shoots.

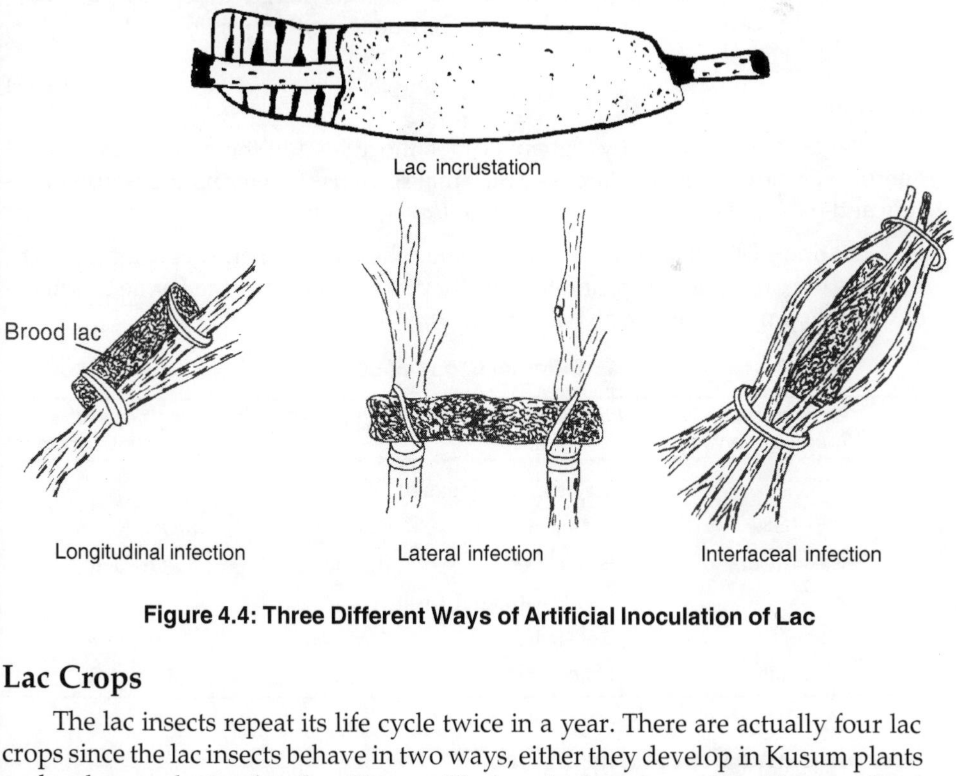

Figure 4.4: Three Different Ways of Artificial Inoculation of Lac

Lac Crops

The lac insects repeat its life cycle twice in a year. There are actually four lac crops since the lac insects behave in two ways, either they develop in Kusum plants or develop on plants other than Kusum. The lac which grows on Non-Kusum plants are called as "Ranjeeni lac", and which grows on Kusum plant is called as "Kusumi lac".

Four lac crops have been named after four hindi months in which they are cut from the tree. They are as follows:

Ranjeeni Crop

(*i*) Katki

Lac larvae are inoculated in June–July. Male insect emerges in August–September. Female give rise to swarming larvae in October–November and the crop is reaped in *Kartik* (October–November).

(*ii*) Baisakhi

Larvae produced by Katki crop are inoculated in October–November, male insects emerges in February–March, females give rise to swarming larvae in June–July, the crop is reaped in *Baisakh* (April–May).

Kusumi Crop

(*i*) Aghani

Lac larvae are inoculated in June-July, male insect emerges in September, female give rise to swarming larvae in January-February and crop is reaped in *Aghani* (December–January).

(*ii*) Jethoi

The larvae produced by Aghani crop is inoculated in the month of January–February, male emerges in March–April, female give rise to swarming larvae in June–July and the crop is reaped in the month of *Jeath* (June–July).

The time of infection with swarming larvae, the time of emergence of male insects, the time of reaping the crop, and the time of producing swarming larvae by female etc., are shown in tabular form below:

Different Crops of Lac

Infection with Swarming Larvae	*Emergence of Male Insect*	*Crop reaped*	*Female give Rise to Swarming Larvae*
	Rajneeni or Nonkusumi Crop		
Katki	August	Oct.-Nov.	Oct.-Nov.
Baisakhi (Oct.-Nov.)	Feb.-March	April-May	June-July
	Kusumi Crop		
Aghani (June-July)	September	Dec.-Jan.	Jan.-Feb.
Jethoi (January)	March-April	June-July	June-July

Scraping and Processing of Lac

Lac cut from the host plant is called as "stick lac". Lac can be scraped from the twigs before or after the emergence of larvae. If it is used for manufacturing before the emergence of larvae, the type of lac produced is called as "Ari lac", and if it is used for manufacturing purpose after swarming of larvae has occurred, the lac is said to be "Phunki-lac".

The scraping of lac from twig is done by knife, after which they should not be exposed to sun. The scraped lac is grinded in hard stone mills. The unnecessary materials are sorted out. In order to remove the finer particles of dirt and colour, this lac is washed repeatedly with cold water. Now at this stage it is called as "Seed lac" and is exposed to sun for drying. Seed lac is now subjected to the melting process. The melted lac is sieved through cloth and is given the final shape by molding. The final form of lac is called "Shellac". Colour or different chemicals may be mixed during melting process for particular need.

Natural Enemies and their control

Natural enemies imposes a challenge to the lac culturist, as they not only decreases the population of lac insects, but also retard the production and quality of lac. Damage caused to lac insects may be grouped under two heads, (a) damage caused by insects (b) damage caused by animals other than insects.

Insect enemies of lac crop may be predators and parasites. The common parasites of lac insect are known as "Chalcid". They are small, winged insects which lay their eggs inside the lac coat either on the body of the lac insect or inside the body of the lac insect. The larvae which hatches from these eggs feed upon the lac insects, thereby causing mortality of their host. Damage done by this parasite constitute about 5-10 per cent of the total destruction of the lac crop.

Damage done by the predators are of greater intensity (35 per cent of the total destruction). The major predators of lac insects are *Eublemna amabilis* (the white moth) and *Holococera pulverea* (the blackish grey moth). They not only feed on lac insects but also destroy the lac produced by them. Squirrels, monkey, rat, bat, birds (wood peckers), man etc., are the enemies other than insects which destruct the lac crop in different ways. Damage is also done by climatic factors such as excess heat, excess cold, heavy rain, storm and partly by the faulty cultivation methods.

Control

Damage caused by the above mentioned animals can be reduced to certain extent by the use of the following methods.

Cultural Method

The amount of damage by infection can be reduced to a greater extent by taking care during the culture of lac insects, especially at the time of inoculation. The brood lac showing the minimum enemy attack should be selected for inoculation and should be cut from the host plant very near to the time of emergence of larvae (about one week before the emergence). This will reduce the chances of parasite attack on the emerging larvae at new place (host). The brood lac used for inoculation should be removed from the new host's branches as soon as the emergence of larvae stops (Approx. 3 weeks after inoculation). It reduces the chance of transference of enemies to the new host plant from the brood lac. The infected brood lac, not fit for inoculation or the used up brood lac should not be retained for long. The lac should be scraped at once and the rest may be crushed or dropped into fire in order to destroy the predators and parasites. The delay in processing also gives chances to the enemy insects to escape into field. So the manufacturers should try to convert stick lac into seed lac as soon as possible. By these cultural methods the future production can be saved from infection to some extent.

Artificial Method

During the crop reaping, it is not always possible for the manufacturers to convert the huge amount of stick lac to seed lac at a time. To avoid the spreading of enemies at this time from stocked stick lac simple artificial method can be used. Bundles of stick lac should be tied with stones and immersed in fenced water (river or ponds) for about a week. This kill all the parasitic and predator insects as they can not survive in water.

Biological Method

It is an indirect method for killing the parasitic and predator insects. For this purpose hyper-parasitic insects are used which attacks the parasitic insects of lac and kill them. These hyper-parasitic insects are however, not harmful for lac crops.

Use of Lac

Lac has been used for the welfare of human beings from the great olden days. No doubt the development of many synthetic produces have made its importance to a little lesser degree, but still it can be included in the list of necessary articles.

Lac is used in making toys, bracelets, sealing wax, gramophone records etc. It is also used in making grinding stones, for filling ornaments, for manufacturing of varnishes and paints, for silvering the back of mirror, for encasing cable wires etc., waste materials produced during the process of stick lac is used for dying purpose. Nail polish is a good example of the by-product of lac.

5

Other Useful Insects

Insect as Pollinator

Pollen is made by the male organs of a plant (anthers in flowers) and contains genetic information needed for plant reproduction. Pollination is the delivery of pollen to the female organs of a plant (stamens in flowers). Pollen may be transferred to female organs on the same plant (self-pollination) or another plant of the same species (cross-pollination). As a result of pollination the plants produce seeds. Pollen can be dispersed by wind, water and animal pollinators such as insects, bats and birds. It is estimated that 65 per cent of all flowering plants and some seed plants (*e.g.* cycads and pines) require insects for pollination. This percentage is even greater for economically important crops that provide fruits, vegetables, textile-related fibres and medicinal products. Because insects are such efficient pollinators, plants have developed many ways of encouraging them to visit. This has led to some strong associations between plants and insects.

Why is Pollination by Insects Important

1. Pollination by insects is a much more reliable and efficient pollination mechanism than chance dispersal.
2. Pollination by insects determines plant community structures.
3. Pollination by insects is particularly important for Australian native trees and shrubs. For example, native bees pollinate many members of the plant family Myrtaceae. This plant family includes eucalypts, angophoras and tea trees.
4. Pollination by insects is vital for crop production. One third of the human food supply is crops that are dependent on pollination by bees.

Which Insects are Pollinators

Species of bees, beetles, flies, wasps, thrips, butterflies and moths are all successful pollinators.

These insects make good pollinators because they share two important features:

1. They fly, and so are capable of visiting many plants in a relatively short amount of time
2. They are motivated to interact with pollen, as they either eat it or food items located nearby *e.g.* nectar.

The most sophisticated relationships between plants and insects are generally those involving bees. Bees collect pollen and nectar not only for themselves but also to feed their young. For this reason bees have developed a number of adaptations that make them particularly good pollen carriers. Bees have special hairs that are arranged to form pollen 'baskets' on their hind legs and the underside of their abdomen. These adaptations allow them to gather and carry large volumes of pollen. Bees are ideal pollinators because they visit many flowers while carrying lots of pollens, before returning to their nest. So the chance that a bee will transfer the pollen between flowers of the same species is very high.

How Insects Pollinate Plants

Pollination by Pollen-feeders

Many insects eat pollen. In the process of eating they become covered in it. Pollination happens when the pollen feeder transfers the pollen to the pollen receivers of the same plant, or another plant of the same species, as the insect looks for more pollen to eat.

Disadvantages of attracting a pollen feeder:

1. They eat the very item the plant wants delivered–pollen.
2. They tend to be generalist feeders and eat other parts of the plant, including the sexual organs.
3. They could be considered to be 'unreliable pollinators' as the pollinator might not go anywhere near the female organs of the same species of plant.

Pollination by Nectar Feeders

The majority of flowering plants encourage insects to visit their flowers by secreting a sugar-rich liquid called nectar. This nectar collects in pools, below the sexual organs of the plant. As the insect enters the flower in search of nectar it brushes against the anthers (pollen bearing male parts of the flower). In doing so the insect collects the pollen, as it sticks to its body. When the insect visits another flower for more nectar, the pollen is transferred from its body to the stamen (pollen receiving female parts of the flower), causing pollination.

Pollination by a nectar feeder has a number of possible advantages including:

1. The locality of the nectar ensures the insect cannot avoid touching the organs associated with pollination.
2. Pure nectar feeders such as butterflies and moths do not eat the pollen.

Effect of Honeybee Pollination on Vegetable Yield

Vegetable	*Per cent increase of seed*
Radish	22-100
Turnip	100-300
Carrot	100-125
Onion	353.5
Muskmelon	756-6700
Cucumber	21.2-411
Squash	771.4-800

Insect as Medicine

There were 21 insects described as medicine in Shennong Pharmacopoeia (100-200AD). This was expanded to 73 insects in Compendium Materia Media published in 1578. Chinese galls (Wubeizi) are perhaps the most common insect-related medicine, used for many sores.They are produced by gall- making aphid (Pemphigidae) on Chinese sumac (Anacradiaceae: Rhus). The caterpillar fungus consists of larvae of *Hepialus armoricanus* (Lepidoptera: Hepialidae) infected with an obligate entomopathogenic fungus *Cordyceps sinensis* (Clavicipitales, Ascomycotina). The pharmacological properties of the caterpillar fungus are similar to these of gingseng. A Chinese medicine book says they are mainly used for weak lungs, coughing and shortness of breath, weak kidney, back pain, impotence etc. Other common insect-related medicines are egg cases of praying mantis and blister beetles. Cicada exuvia they are supposely good for scrofula and ulcer. Silkworm frass is also used as a medicine for diarrhea.Cockroach (*Eupolyphaga sinesis*) is also used as a medicine. It is supposed to help stop bleeding and heal bone fractures, swelling etc.

Apitherapy is now very popular in China. Arthritis is the most popular disease to be treated by bee stings. Recently, apitherapy is also used for both arthritis and for muscular dystrophy in US. Other bee products used for medicine include honey, propolis (used in an alcoholic tonic). Royal jelly is very popular as a health-strengthening food, especially among the "intellectualls". Use of pollen as health food is rather recent, perhaps after the Europeans. Queen larvae, mostly by-products of royal jelly production, are also used for making alcoholic tonic.

Ants are used as health food and a medicine, though as usual it is not known what are the active ingredients. There had been an anecdotal report that a village of long-living people (average ~90) attributed their longevity to the habit of frying up ants and eating them. Ant is a major component of a herbal medicine for hepatitis B. This medicine is reported to give a 60 per ecnt efficiency to convert hepatitis B surface antigen (HBsAg) to serum negative. This compares favorably with the ~30 per cent conversion efficiency using interferon, as reported by medical journals in US. There are also wines and tonic made with ants.

Insects as Food

Nymphs of cicadas, larvae/pupae of wasps and ants were used as delicacies for emperors and nobles. Locusts are also used as food, in many parts of world. Pupae of silkworms which just finished producing silk was also good food. Predacious diving beetle and giant water-bug are popular in China. Dragon flies are caught and grilled, in China, Japan, Laos and Thailand. Entomophagy (the eating of insects) has yet to become a day-to-day activity for many people in world. In spite of the superior nutritional content of edible insects compared to other animals. Other cultures around the world have made insects a main ingredient in their diets, providing an excellent source of protein. Insects are an inexpensive substitute for meat in many developing countries. In Mexico, grasshoppers and other edible insects are sold by the pound in village markets and are fried before being eaten. Many are sold in cans as fried grasshoppers, chocolate covered ants, etc. Columbian citizens enjoy eating a variety of insects such as termites, palm grubs and ants. Ants are ground up and used as a spread on breads. Popular insects eaten in the Phillippines are June beetles, grasshoppers, ants, mole crickets, water beetles, katydids, locusts and dragonfly larvae. They can be fried, broiled or sauteed with vegetables. In parts of Africa, ants, termites, beetle grubs, caterpillars and grasshoppers are eaten. Some insects such as termites are eaten raw soon after catching, while others are baked or fried before eating. The giant waterbug roasted and eaten whole is a favorite food in Asia. It is easily collected around lights at night around bodies of water. Sago grubs are popular for cooks in Papua New Guinea, most often boiled or roasted over an open fire. Other edible insects eaten include larvae of moths, wasps, butterflies, dragonflies, beetles, adult grasshoppers, cicadas, stick insects, moths and crickets.

Insect as Biocontrol Agents

Fifteen of the 29 orders of insects have species that naturally parasitise or prey on other organisms that humans classify as detrimental. Hymenoptera and Diptera contain the greatest number. These bio control agents are very useful in insect pest and weed management programme.

Some Important Insect Parasitoids and Predators of Crop Pests and Weeds

Sl.No.	*Parasitoids/Predators*	*Used Against*
1.	*Trichogramma* spp.	Egg parasitoid of moths
2.	*Bracon hebetor*	Larval parasitoid
3.	*Chelonus blackburnii*	Egg-larval parasitoid
4.	*Goniozus nephantidis*	Larval or pre-pupal ectoparasitoid
5.	*Chrysoperla carnea*	Predators of soft bodied insects
6.	*Cryptolaemus montrouzieri*	Predators of mealy bugs

6

Polyphagous Pests

Locust

Locusts are members of the grasshopper family acrididae which included most of the short-horned grasshopper.

Types of Locust

There are some 5000 species of grasshoppers found in the world, but only 9 are recognized as locust of which 3 belong to the Indian subcontinent. These are as follows:

1. *Migratory locust (Locusta migratoria)*–Occurs in India, Europe, Africa, Sri Lanka, Pakistan, East and South Asia, Australia.
2. *Bombay locust (Patanga succinacta)*–Occures in India, Sri Lanka, Malaysia.
3. *Desert locust (Schistocerca gregaria)*–Occurs in India, Pakistan, Arabia, North African countries.

Desert Locust

The desert locust is the most destructive of all locusts. It invades 30 million square kilometer areas spreading over 60 countries from the west and north Africa to Assam in India and in 50 per cent of this area, breeding can occur.

Distribution of Desert Locust

The desert locust is an inhabitant of the dry grasslands of desert areas and is found in many countries of the world. Its distribution extends from Pakistan to Afghanistan, Iran, Iraq, Arabia and Northern Africa. In India breeding grounds are located in Rajasthan, part of Gujarat and the Hisar and Mohindergarh district of

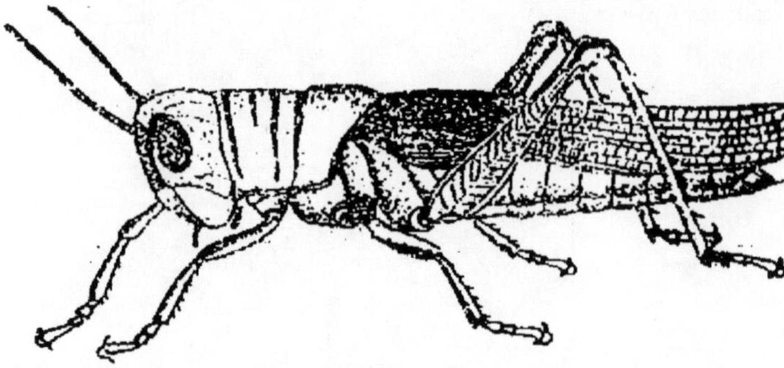

Figure 6.1: Desert Locust

Haryana. These places are not the permanent home but are merely the outbreak areas where the locusts undergo change in their phase from gregarious to solitary.

Phases of Locusts

The desert locust is found in two phases:

1. *The solitarious phase*: When individuals are at low densities.
2. *The gregarious phase*: When they are at high densities.

The transition from the solitarious phase to gregarious and vice versa is called the transient phase.

Difference Between Solitarious and Gregarious Phases

The characteristics of these phases are different from each other, particularly in the colour of their nymphs.

Solitarious Phase	*Gregarious Phase*
1. The nymphs (hoppers) are varied in colour according to the colour of the surrounding vegetation.	1. The nymphs are yellow or pink with distinct black markings.
2. The adult remain greenish grey throughout their life time.	2. The adults are pink on emergence, gradually turning grey and ultimately yellow, when sexually mature.
3. Numbers of molts are 5 to 6 (nymphs)	3. Number of mollts are 5 times (nymphs)
4. Hoppers remain scattered on vegetation whereas adults during nights as isolated individuals.	4. Hoppers form groups or band and march long distances, whereas, adults generally fly during day times in swarms.

The Life Cycle

The desert locust like all other locusts and grasshoppers passes through three stages: egg, nymphs (hoppers) and adults.

Oviposition and Incubation

The female lay eggs in batches called egg-pods, they look like rice grains and are arranged like a miniature hands of bananas. The female bores into the ground with the valves at the rear of the abdomen and deposits a batch of eggs. She then fills up the hole above the eggs with a plug of froth. The pod is about 5 cm long and is laid with its top 5-10 cm below the surface. This is a surprising depth which requires a great extension of the abdomen. The desert locust lays pods containing less than 80 eggs in the gregarious phase and between 95 to 158 in the solitarious phase. The eggs are initially laid in bare ground and often, but not exclusively in sandy soil. As a rule, the female will not lay unless soil is moist at about 5-10 cm below the surface. In soft sandy soils, females have been known to lay when soil moisture is only found at depths below 12 cm. The number of egg pods a female lays depends on how long it takes for her to develop a pod and how long the female lives.

Hopper Development and Behaviour

After hatching, the emerging hoppers work their ways up the forth plug to the surface. They immediately moult to the first instar. The hoppers then pass through five instars shedding a skin (moulting) between each. At the first moult (called fledgling) the young adult, known as a fledgling emerges.

Adult Development and Behaviour

It takes about ten days after fledging for the adult wings to become hard so that it is capable of sustained flight. The adult then remain immature until they encounter conditions which stimulate maturation. There is however, a maximum period. Adults in an area of lush vegetation with maximum day temperature of 35°C or more, and with rain to maintain the vegetation growth, can probably lay within three weeks of fledging. At the other extreme, immature adults, can survive for six months or more under dry conditions. Adults cannot survive long under hot dry conditions with little to eat, migrations to areas where the rainfall is one way in which plagues collapse. Adults can also survive during winter in West Africa South of the Sahara where it is relatively warm, but these adults do not breed.

Breeding Season and Migration of Swarms

The breeding of locust depends upon the rainfall and the subsequent vegetation. The eggs are laid in the sandy soil and adequate moisture is required before they can hatch. In India, there are two breeding seasons during the year, *viz.*, (1) The summer breeding season and (2) The monsoon breeding season.

Origin of New Locust Cycle

The following sequence of events generally mark the beginning of the locust cycle: (*a*) extensive breeding on coastal areas of Arabian countries as a result of heavy winter and spring rainfall and the formation of gregarious swarms. (*b*) migration of locusts from Arabian coast into the interior (Baluchistan, Afghanistan) in spring. (*c*) migration of these swarms in summer, to Sind and Rajasthan. (*d*) extensive breeding of these swarms in Rajasthan in July–August and September. (*e*) migration of the subsidiary gregarious swarms from Rajasthan to Punjab in late summer.

Development of Desert Locust in Various Broods in India

Sl.No.	Stage of Development	Summer Brood	Winter Brood
1.	Pairing and eggs laying	February–April	July–September
2.	Hoppers	March-May	July–October
3.	Adults	May–June	September–October

Management

1. The adults can be beaten to death with thorny stick, brooms or can be swept together and buried underground in heaps.
2. Methyl parathion (5 per cent) or endosulfan (2 per cent) if dusted on crops, trees and the ground, is very effective.
3. If eggs are laid in well defined area, a trench may be dug around it, so that the young nymphs on emerging drop into it and can be buried alive, filling the ditch with soil.
4. At night when the hoppers rest on bushes, they can be burnt with flame-throwers.
5. Poison bait such as the poisoned bran mash or sawdust, if scattered in the early morning or in the evening, are effective. The poison used is a sodium fluosilicate or paris green.
6. A number of birds attack locusts and of these, the common *myna* and the *tiliar* (starling) are the most important.

Termite or White Ants, *Odentotermes obesus, O. assumthi, Microtermes obesi, Coptotermes heimii* (Isoptera: Termitidae)

Host Range

It is polyphagous pest and one of the most destructive creature of all cultivated crop plants throughout the world.

Marks of Identifications

The termites are social insects and their colony organization is based on caste system.

(A) Reproductive Castes

(i) Winged Male and Female

Winged individuals of both the sexes are produced in large numbers during the rainy seasons. They are brownish in colour with two pairs of slender, dark brown, long narrow wings which are used for nuptial flight only and when they have mated, the wings are then drop off.

(ii) Queen

The queen is wingless, creamy white in colour, abdomen marked with transverse dark brown strips, 60-80 mm in length and 10 mm in thickness. She is a mother of the colony and lives in a specially prepared royal chamber, which is situated in the centre of the nest.

(iii) King

King is much smaller than queen and remains with queen in royal chamber.

(iv) Supplementary Reproductive

They are short winged or wingless creature of both the sexes. They may be present among the offspring of the primary reproductive and are not normally found in colonies and appear to replace the primary reproductives when the later die.

(B) Sterile Casts

(i) Workers

They are sterile, apterous individuals with non- functional reproductive organs, about 6-8 mm long, pale in colour, brown head and eyes are small or absent. Mandibles are well developed. They work for the whole colony by storing food, looking the young ones and building up the termitarium. In a colony large numbers of workers are present.

(ii) Soldiers

Bigger than workers, sterile, head and thorax well chitined, abdomen is delicate and dirty white in colour. They are 5-10 per cent in number in a colony. They are winged. The main function of soldiers is to protect the colony.

Nature of Damage

Termite feed on all the things which contain cellulose. In loamy or light soil and dry areas where proper facilities for irrigation are not usually available, the termite infestation is more serious. The infestation of termite is more on *rabi* crops as compares to *kharif* crops. It feeds on the roots of all the vegetable plants. As a result of infestation, the leaves of infested plants starts drying, such plant can easily be pulled out from the soil. In later stage the whole plants are withered and dry.

Life Cycle

In the rainy season, when the weather conditions are favourable, the colonizing forms emerge from their nest and fly in swarms for a little time. The swarm of colonizing individuals has members of both the sexes. Swarming usually takes place in the day time and most of the individuals of the swarms are destroyed by birds, etc. the survival mate, shed their wings and burrow in the ground to form a new colony of which they become the king and queen. Seven to ten days after swarming the female lays the first batch of 100-130 eggs. The queen can lay upto 30,000 eggs per day. Eggs are kidney shaped and light yellow in colour. The incubation period is about one week in summer. Within 6 weeks the larvae develop to form soldiers or workers as per there requirement of the colony. The reproductive castes mature in 1-2 years. There is only one queen in the colony and normally her longevity is for 5-10 years. The king's life is much shorter than that of queen and when he dies, is replace by new one (supplementary reproductives). Only one generation is completed in a year.

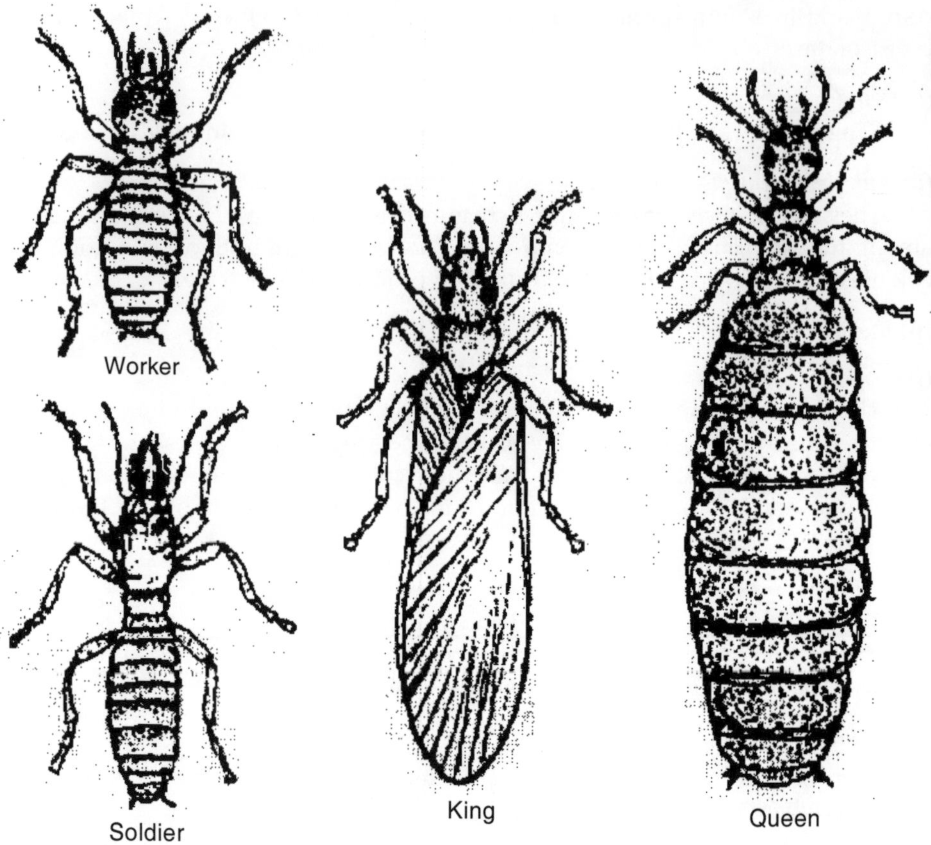

Figure 6.2: Termite

Management

1. Use well decomposed organic manure. Never use raw manure or cow dung. Remove dead or decaying matter or dry stubbles from the field to avoid termite infestation.
2. It is seen that irrigation is useful to protect the crop from termite.
3. Treat the soil with quinalphos 1.5 per cent or methyl parathion 2 per cent dust @ 25 kg per ha before sowing or planting the crop or seed treatment with chlorpyriphos 20 EC @ 6 ml per kg seed or endosulfan 35 EC @ 6 ml per kg seed has also been recommended for effective protection of various crops.
4. In standing crop apply chlorpyriphos 20 EC @ 4 litre per ha with irrigation water for effective protection of crop against termite infestation.

7
Pest of Cereals

Pest of Cereals

Rice

Brown Planthopper, *Nilaparvata lugens* (Hemiptera: Delphacidae)

Marks of Identifications

Adult hopper is 4.5-5.0 mm long and has a yellowish brown to dark brown body. The wings are sub hyaline with a dull yellowish tint. It has two characteristic wing morphs: macropterous (long winged) and brachypterous (short winged). (Wing morphism is influenced by various factors *viz.*, crowding during the nymphal stage and reduction in the quality and quantity of food, short day length and low temperature, which favour macroptery).

Nature of Damage

Both the nymphs and adults remain at the ground level and suck the plant sap. It is a typical vascular feeder primarily sucking phloem sap leading to hopper burn. At early infestation, circular yellow patches appear which soon turn brownish due to the drying up of the plants. The patches of infestation then may spread out and cover the entire field. The grain setting is also affected to a great extent. During sustained feeding, it excretes a large amount of honeydew. It also acts as vector of the virus diseases like grassy stunt, wilted stunt and ragged stunt. (Transmission of persistent ragged stunt and grassy stunt virus require more time. Sheath blight and stem rot incidence was high in BPH infested plants). Symptoms will not be visible from outside in the early stages, but if we enter the field and tap the plants large number of this insect can be seen. They are visible only when the damage has been severe, the plants present a burnt up appearance, hopper burn, in circular patches.

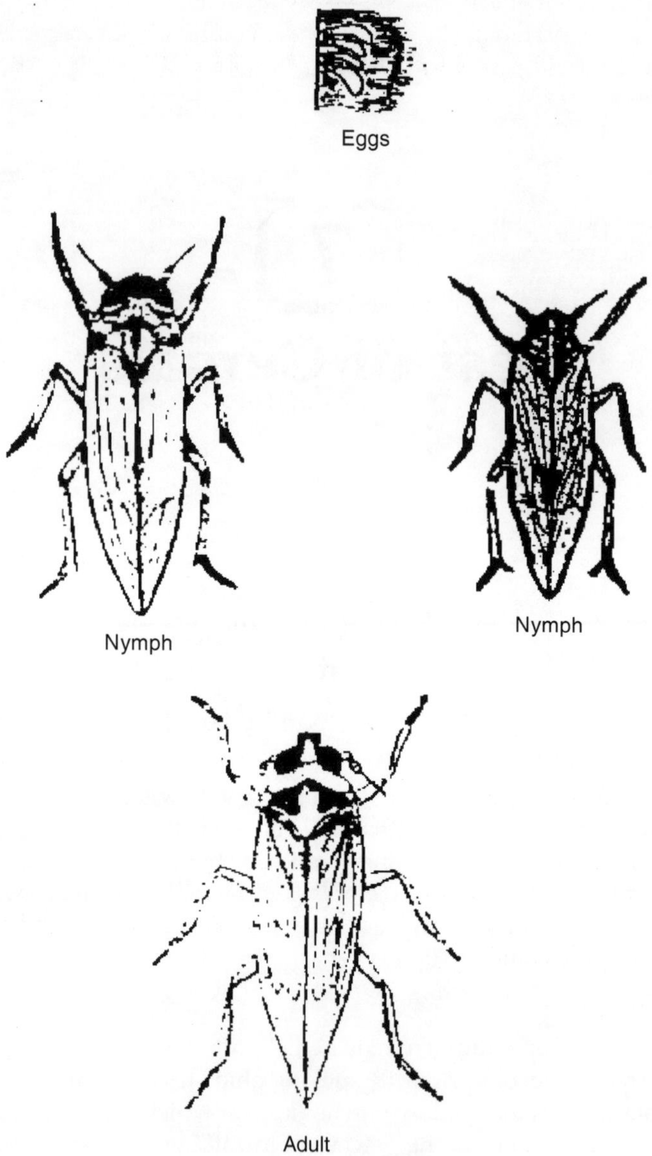

Figure 7.1: Brown Plant Hopper

Life Cycle

Adult female lays eggs in groups towards the lower side of the plant. It makes an incision and inserts the eggs inside the leaf sheath. The incision can be marked as a brownish patch from out side of the stem. Each female lays about 100-200 eggs during its life span. Eggs hatch after 7-10 days after oviposition and the 1st instar nymph comes out. The insect has five nymphal instars during its developmental period.

Nymphal development takes about 2-14 days to become adult. Two forms of adult stages in both male and female insects have been found i. e. macropterous (winged and brachypterous (wingless) forms. The adult female BPH starts egg laying after 2-3 days of emergence. When the population of insect in a area becomes more, macropterous forms migrate and infest new crop areas.

Management

1. Adopt planting with formation of alleys of 25 cm at intervals of 2 mts to provide good aeration and sunlight.
2. Avoid dense planting, planting 33 hills in kharif and 44 hills in rabi per sq. m. may be followed.
3. Excess application of N fertilizers may be avoided.
4. In vegetative phase of the crop growth periodical drying and wetting may be followed for short period to create disturbance in micro climatic conditions favorable to pest development.
5. Grow resistant varieties *viz.*, Chaitanya, Krishnaveni, Chandan, Triguna, Deepthi, Nandi, Vijeta,Pratitha, Vajram, etc.
6. Egg parasites like *Angrus* spp and nymphal and adult parasites like *Pseudogonatonus* spp., were observed to exercise control to the extent of 10 to 40 per cent. Mirid bugs, *Cyrtorlinus livipennis*, is one of the most important predator of BPH in rice ecosystem.
7. Application of Carbofuran 3 G @ 25 kg/ha or spraying of ethofenprox 10 EC at 2 ml or chlorpyriphos 20 EC at 3 ml per litre of water found effective.

Green Leafhopper, *Nephotettix virescens*, (Hemiptera: Cicadellidae)

Marks of Identifications

Adults are 3-5 mm long, bright green with variable black markings, wedge shaped with a characteristic diagonal movement. Male insect has a black spot in middle of the forewings that is absent in females. The insect is active during July to September. The nymphs are soft bodied, yellow white in colour. Gradually the colour changes to green.

Nature of Damage

Both nymphs and adults suck the plant sap from the leaf and leaf sheath. (It is a phloem feeder. Amino acid content is high in phloem sap than xylem. The xylem and phloem vessels are plugged with their stylet sheath that causes disruption in the transport of food substances in the vessels.) Mild infestation reduces the vigour of the plant and the number of reproductive tillers. Heavy infestation causes withering and complete drying of the crop. Plants are predisposed to fungal and bacterial infection through feeding and ovipositional punctures. Nymphs and adults exude sticky, whitish honeydew, which attracts sooty mould (that reduces the photosynthetic rate). It also transmits plant diseases such as dwarf, transitory yellowing, yellow dwarf and rice tungro virus (Tungro is transmitted during short feeding period).

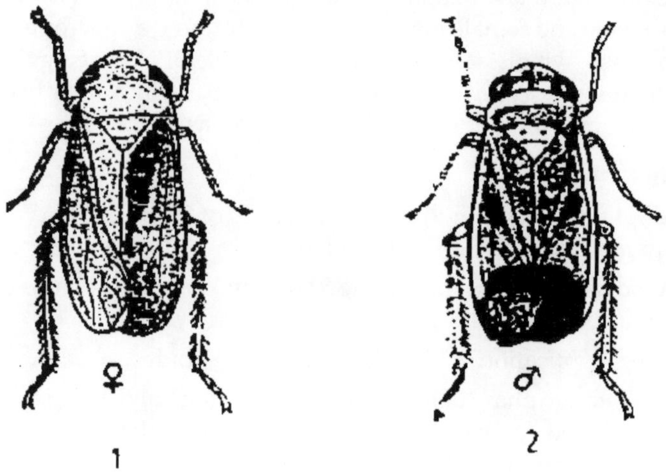

Figure 7.2: Green Leaf Hopper

Affected plants become pale yellow in colour and get stunted in growth. If the plants are tapped large number of leafhoppers may be seen jumping to water.

Life Cycle

The female lay eggs on the inner surface of the leaf-sheath in groups of 3-18. The eggs hatch in 3-5 days and the nymphal stage is completed in 12-21 days. The adult live for 7-22 days. There are 6 generations in a year. The insect overwinters in the adult stage.

Management

1. Adopt planting with formation of alleys of 25 cm at intervals of 2 mts to provide good aeration and sunlight.
2. Avoid dense planting, planting 33 hills in kharif and 44 hills in rabi per sq. m. may be followed.
3. Excess application of N fertilizers may be avoided.
4. In vegetative phase of the crop growth periodical drying and wetting may be followed for short period to create disturbance in micro climatic conditions favorable to pest development.
5. Grow resistant varieties *viz.*, Chaitanya, Krishnaveni, Chandan, Triguna, Deepthi, Nandi, Vijeta,Pratitha, Vajram, etc.
6. Egg parasites like *Angrus* spp and nymphal and adult parasites like *Pseudogonatonus* spp., were observed to exercise control to the extent of 10 to 40 per cent. Mirid bugs, *Cyrtorlinus livipennis*, is one of the most important predator of BPH in rice ecosystem.
7. Application of carbofuran 3 G @ 25 kg/ha or spraying of ethofenprox 10 EC at 2 ml or chlorpyriphos 20 EC at 3 ml per litre of water found effective.

Whitebacked Planthopper, *Sogatella furcifera*, (Hemiptera: Delphacidae)

Marks of Identifications

The adult hopper is 3.5-4.0 mm long. The forewings are uniformly hyaline with dark veins. There is a prominent white band between the junctures of the wings. Macropterous males and females and brachypterous females are commonly found in the field.

Nature of Damage

WBPH is more abundant during the early stage of the growth of rice crop, especially in nurseries. (It attacks less than four-month old plants in fields with standing water and shows a marked increase with the age of the crop. Rice is more sensitive to attack at the tillering phase than at the boot and heading stages.) Damage is caused through feeding and oviposition. Gravid females cause ovipositional punctures in leaf sheaths. Both nymphs and adults suck phloem sap causing reduced vigour, stunting, yellowing of leaves and delayed tillering and grain formation (Rice crop fails to produce complete grains [seedless glumes] and this condition is known as red disease in Malaysia). Feeding puncture and lacerations caused by ovipositor predispose the plants to pathogenic organisms and honeydew excretion encourages the growth of sooty mould. It is not a vector of any viral disease.Heavy infestation cause outer leaves of a hill to show burn symptoms. Damage in the form of hopperburn appears uniformly in a rice field, whereas it appears as circular patches in the case of BPH.

Life Cycle

Pre mating period is 2 days, pre-oviposition and oviposition periods are 2 and 45 days, respectively. Eggs are laid in clusters inside the leaf sheath after injuring the surface with the ovipositor. A female lays about 85 eggs masses and about 546 eggs. The incubation period is 7-8 days. Nymphal period is 15 days. There are 5 nymphal instars. Male longevity is 40 days, whereas female survives up to 50 days. Total life cycle is 25 days (egg to egg).

Management

1. Timely planting, and judicious use of nitrogenous fertilizers.
2. Grow resistant varieties like Tulasi, Tripti, Salivahan, Haryana Basmathi-1 and IR-62.
3. The following spiders and insects are potential predator of the WBPH. Wolf spider (*Lycosa chaperi*), jumping spider (*Plexippus paykulli*), orb weaver or garden spider (*Araneus sinhagrdensis*), crab spider (*Thomisus cherapunjeus*), four jawed spider (*Tetragnatha mandibulata*), lynx spider (*Oxyopes pandae*), ground beetle (*Casnoidea indica*), ladybird beetle (*Coccinella arcuata*), etc. Encourage their activities in the field.
4. Spray chlorpyriphos or monocrotophos @ 0.5 per cent a/i. per ha or carbofuran @ 1.0 kg a. i. per ha, can control the pest effectively.

Gundhi Bug, *Leptocorisa acuta*, (Hemiptera: Coreidae)

Marks of Identifications

Adults are greenish yellow, long and slender, above ½ inch in length with a characteristic buggy odour.

Nature of Damage

Both adults and nymphs do the damage. The nymphs start feeding 3 to 4 hours after hatching. They feed on the leaf sap near the tip/on milky sap in developing spikelets at milky stage. Sucking of the milky sap causes ill-filled/partial filled and chaffy grains. Serious infestation can reduce the yield by 50 per cent. The straw gives off-flavour that is unattractive to cattle. Leaves turn yellow and later rusted from tip

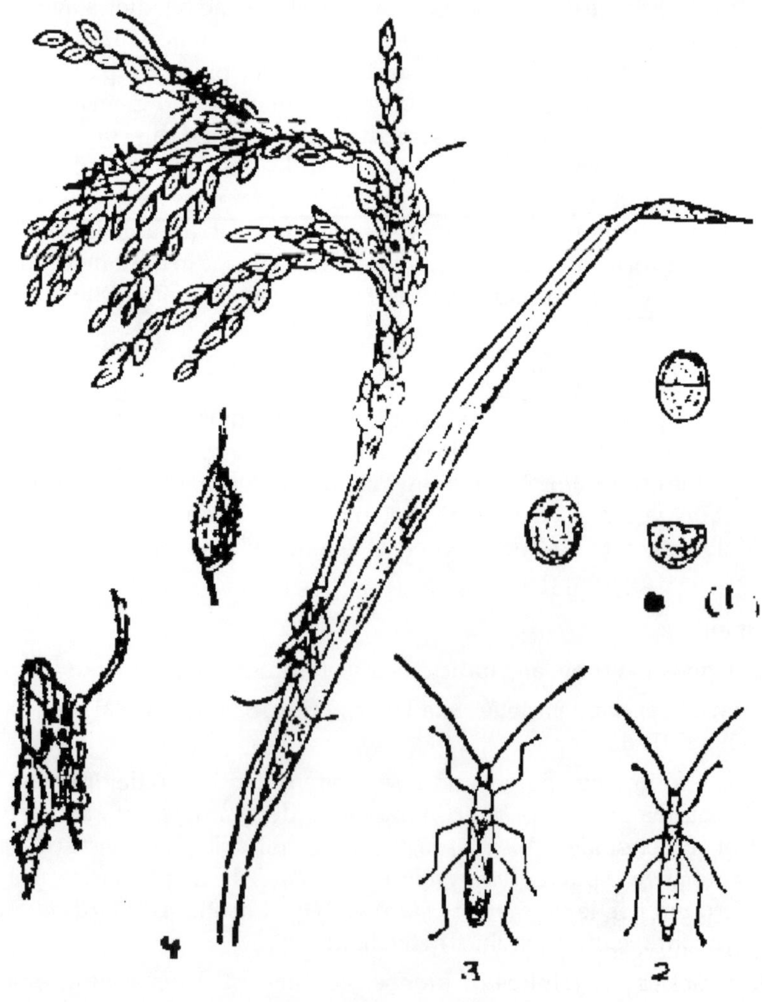

Figure 7.3: Gundhi Bug

downwards. Appearance of numerous brownish spots at the feeding sites/shrivelling of grains. In the case of heavy infestation, the whole earhead may become devoid of mature grains. Its presence in the field is made out by its strong smell.

Life Cycle

The pest breed all the year on main crop and grasses. The female lay 24-30 round yellow eggs in row on the leaves. The eggs hatch in about 6-7 days and the nymphs grow to maturity in 6 stages within 2-3 weeks. The adult bug live for 33-35 days. Many generations are completed in a year.

Management

1. Removal of weeds in the vicinity of paddy crop as the pest breeds on a variety of grasses prior to its migration to rice crops.
2. Collection of bugs by hand netting is useful. Sweeping of rice plants with winnows smeared with sticky material lake castor.
3. Spraying of monocrotophos 1.6 ml or endosulfan 2 ml/lit of water once at flowering and another at grain hardening stage or dusting of endosulfan 4 per cent @ 25 kg/ha in evening hours. Repeat this after 10 days if needed.

Paddy Stem Borer, *Scirpophaga (=Tryporyza) incertulus*, (Lepidoptera: Pyralididae)

Marks of Identifications

They exhibit remarkable sexual dimorphism. The female moth is bright yellowish brown with a black spot at the centre of the forewing and a tuft of yellow hairs at the anal region. The male is small in size and brownish.

Nature of Damage

The insect may start attacking the plants in the nursery especially long duration varieties. The incidence is mild in the season June to September, but later on gets intensified from October to January and February. The caterpillar enters the stem and feeds on the growing shoot. As a result the central shoot dries up and produces the characteristic dead heart. The tillers may get affected at different stages. When they are affected at the time of flowering the earheads become chaffy and are known as white ear. A number of stem borer moths seen dead and floating on the water in the fields. In the vegetative stage, dead hearts seen in the affected tillers and in the reproductive stage, white ear may be seen. The full-grown caterpillar measures about 20 mm, white or yellowish white in colour with a conspicuous prothoracic shield.

Life Cycle

The female lay eggs on the under side of the leaves. A single female lay about 120-150 eggs. The eggs are covered with yellowish brown hairs of the female tuft. They hatch in 6-7 days. The larva grow in 6 stages and is full-fed in 16-27 days. It then constructs an emergence hole which always located above the water level and pupates inside the attacked plant. Pupal period lasts 9-12 days. The life cycle is completed in 31-46 days. There are 2-5 generations in a year in different agroclimatic zones.

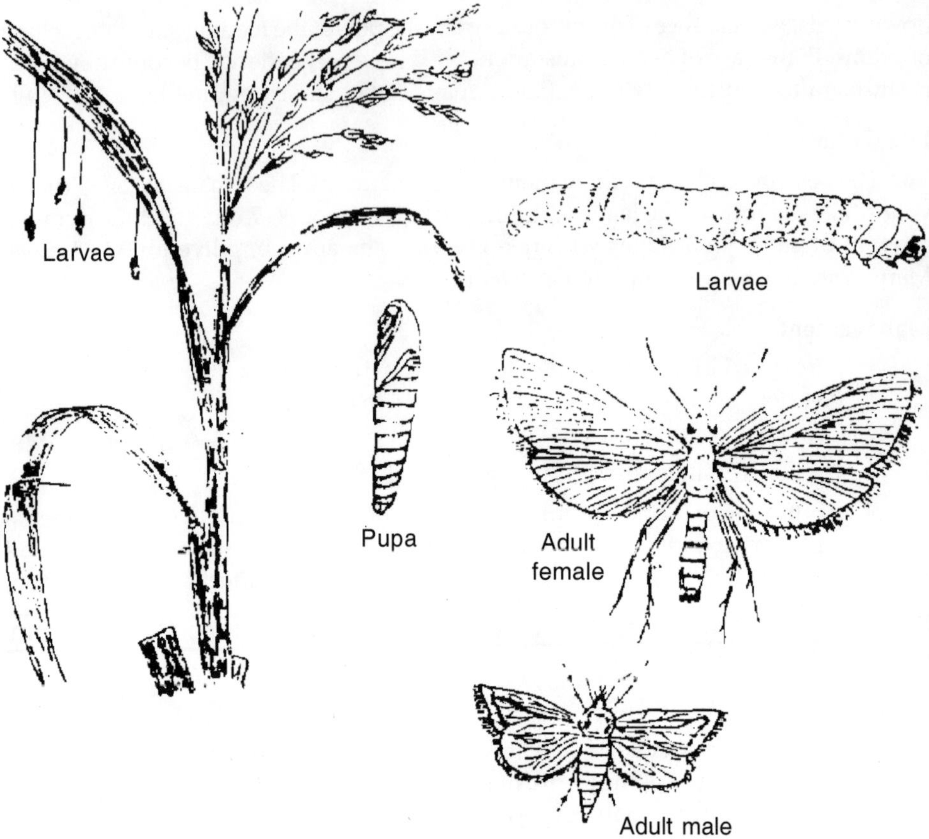

Figure 7.4: Paddy Stem Borer

Management

1. Paddy stalks should be harvested close to the ground.
2. The field should be ploughed in the summer months and the stubbles are to be collected and burnt.
3. Sowing and planting time should be adjusted to avoid peak borer infestation.
4. Crop rotation with pulses, vegetables or groundnut after kharif rice should be followed to prevent continuous build up of borer population.
5. Heavy pasturing of stubble field and ploughing in summer which expose the roots of stubbles also reduce borer population.
6. Since the eggs are laid near the tip of the leaf blade, clipping the seedlings before transplanting reduces the carry-over of eggs from seed bed to the transplanted field. Harvesting at ground level or ploughing after harvesting remove majority of larvae and pupae.
7. The use of sex pheromone traps for mass trapping of male moths.

8. *Telonomus* spp., *Tetrastichus* spp and *Trichogramma* spp are identified as dominant complex stem borer parasitoids, they could be utilized.
9. Sasyasree (RNR 446), Rathna and Kaveri were observed tolerant to this pest in many parts of India.
10. Spray chlorpyriphos 2 ml/lit and repeat the same at 10-15 days. Apply cartap hydrochloride 4 G @ 20 kg/ha or carbofuran 3G @ 25 kg/ha.

Paddy Gall Midge, *Orseolia oryzae,* (Diptera: Cecidomyiidae)

Marks of Identifications

The adult fly is yellowish brown and mosquito like. The male is ash grey in colour. Adults feed on dewdrops.

Nature of Damage

The maggot bores into the growing point of the tiller and causes abnormal growth of the leaf sheath, which becomes whitish tubular and ends bluntly. It may be pale green, pink or purplish. Further growth of tiller is arrested. This is called onion shoot, silver shoot or anaikomban. The feeding by the maggot and the larval secretion, which contains an active substance called cecidogen, is responsible for cell proliferation of the meristematic cells and gall formation. It is a pest in irrigated and wet season crop. Tillers in 35 to 53 days old crops are preferred. The central shoot instead of producing leaf, produces a long tubular structure. When the gall elongates as an external symptoms of damage, the insect will be in pupal stage and ready for emergence.

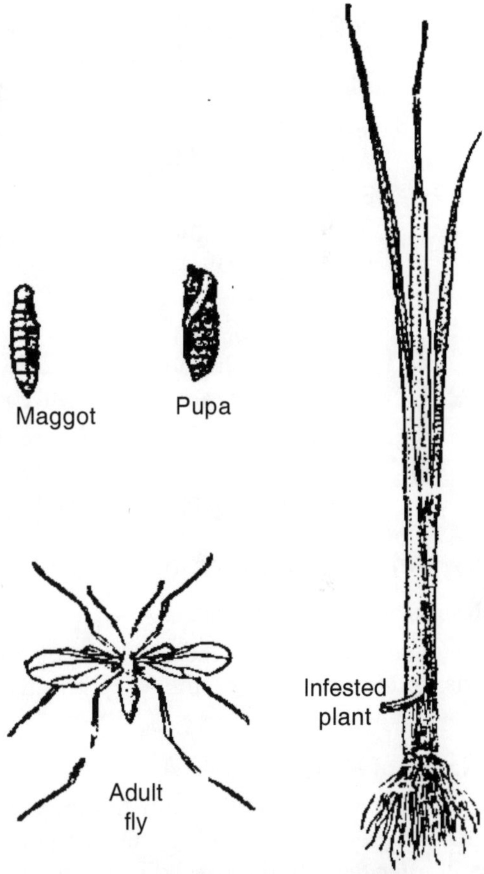

Figure 7.5: Paddy Gall Midge

Life Cycle

Adults are nocturnal, phototropic and live for 2-5 days. Mating usually takes place soon after emergence and oviposition initiates a few hours latter. Female lays 100-200 eggs, either singly or in groups of 3-4 on the ligules or in their vicinity on the leaf blade, or on the leaf sheath. Incubation time is 3-4 days. The newly hatches maggots of both sexes are about 1 mm long and can live in water upto 3 days without any adverse effect. They crawl down the leaf sheath to the growing points of the tillers and reach the interior of the bud, where they lacerate the tissues and feed until pupation. The feeding stimulates tillers to grow into tubular gall that resembles an onion leaf. The average larval period is 15-20 days. The maggot moults thrice at 3, 5 and 7 days for 1st, 2nd and 3rd instars, respectively. Pupation occurs inside the gall near the base. Pupal period is 5-8 days. Life cycle completes in 19-21 days in rainy season but in winter months it takes 32- 39 days. There are 3-5 overlapping generations on the same crop and 5-8 in a year.

Management

1. Advancing the planting date needs to be adopted to avoid the peak activity of the pest to protect the crop. Late and closer planting predispose the crop to gall midge infestation.
2. Off-season activity of gall midge in rice stubbles or alternate weeds like *Echinocloa crus-galli* and *Leersia hexandra* can be checked by removal of these plants.
3. Use resistant varieties *viz.*, Shakti, Samalei, Kakatiya, Surekha, Phalguna and Vikram and Rajendra dhan.
4. Seed treatment with chlorpyriphos or isofenphos (0.2 per cent solution) for 3 hours or seed mixing with chlorpyriphos (0.75 kg a. i. per 100 kg seeds) or imidacloprid (0.05 kg a. i. per 100 kg seeds) provides protection for 30 days in the nursery.
5. Seedling root dip for 12 hours before planting in chlorpyriphos, isofenphos or chlorfenvinphos at 0.02 to 0.04 per cent concentration has provided most effective and economical control.
6. In case of late infestation, based on the economic threshold level effective granular insecticides like phorate, carbofuran or quinalphos can be broadcasted in one or two rounds.

Swarming Caterpillar, *Spodoptera mauritia* (Lepidoptera : Noctuidae)

Marks of Identifications

The adult moth is medium sized, stout built dark brown with a conspicuous triangular black spot on the forewings. Hind wings are brownish white with thin black margins.

Nature of Damage

Caterpillars march in large numbers in the evening hours and feed on the leaves of paddy seedlings till the morning and hide during daytime. They feed gregariously and after feeding the plants in one field march onto the next field. Under severe infestation crop gives the appearance of grazed plants. Attacked plants are reduced to stumps. Nurseries situated in ill-drained marshy areas attacked are earlier than dry ground. Damage is severe during July to September. Nurseries found completely eaten away by the caterpillars' overnight.

Life Cycle

The pest is active throughout the kharif season but its damage is noticed in the early part, especially after good rains when it appears in the form of a true army

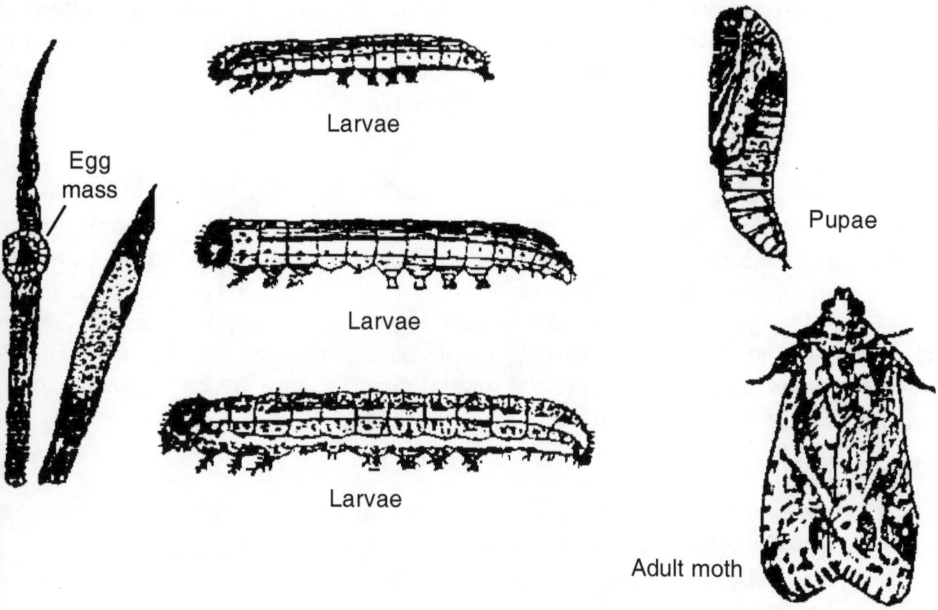

Figure 7.6: Swarming Caterpillars

worm. There may be three to four brood in a season and generally the second brood is most destructive. The female moth is nocturnal in habit and lays just 24 hours after the emergence. After 24 hours of mating egg laying takes place. The eggs are laid in batches on various kinds of wild grasses and paddy leaves. The number of eggs laid per batch may be 200-300. The fully grown larva is smooth, cylindrical and has a large pale colour with sub-dorsal and dorsal strips. The larval period lasts for 30 days and pupal period is about 8-10 days. Adult longevity is about 2-4days.

Management

1. In case of severe infestation the entire area should be isolated by trenching and the crop should be ploughed up.
2. Collection of larvae with a hand net or sweeping basket and their destruction.

Rice Case Worm, *Nymphula depunctali* (Lepidoptera: Pyralididae)

Marks of Identifications

The adult is a small delicate moth having white wings speckled with pale brown wavy markings. Females are larger than males. Egg laying takes place in the night.

Nature of Damage

The caterpillar cuts a piece of leaf, rolls it longitudinally into a tubular structure and remains inside. It feeds by scraping the green tissue of the leaf. The cases often float in the water. Its damage can be distinguished from damage by other pests in two ways, firstly the ladder like appearance of the removed leaf tissue resulting from the back and forth motion of the head during feeding and secondly the damage pattern is not uniform through out the field because the floating cases are often carried in the run off water to low lying fields where the damage is more concentrated. Plants stunted, caterpillars hanging on the leaf edges in a tubular case.

Life Cycle

Eggs are laid on leaves and leaf ssheath in rows and batches. A single female lay upto 150 eggs. The eggs hatch in about one week. Young larvae feed by scrapping the leaf surface. The larvae become full grown in 20 days. The larvae undergoes 6 instars and is characterized by the presence of tubular gills on its body. The larval stage pupate inside the last case. Before pupation the case is attached to the leaf sheath above the water level and its both ends are plugged. The pupal period lasts for about a week. The pest is active during the monsoon and there may be 2-3 broods in a season. The life cycle is completed in about 35-40 days.

Management

1. Draining out the stagnant water from the field is very much effective in reducing the population of the pests.
2. Destruction of the weeds around the field.
3. Introduction of *Elsamus* spp., *Apanteles*, *Bracon*, *Hormius* etc for its effective control.

Rice Hispa, *Dicladispa armigera* (Coleoptera: Chrysomelidae)

Marks of Identification

The adult beetle is somewhat square shaped about 1/6 to 1/8″ in length and width. Dark blue or blackish in colour with spines all over the body.

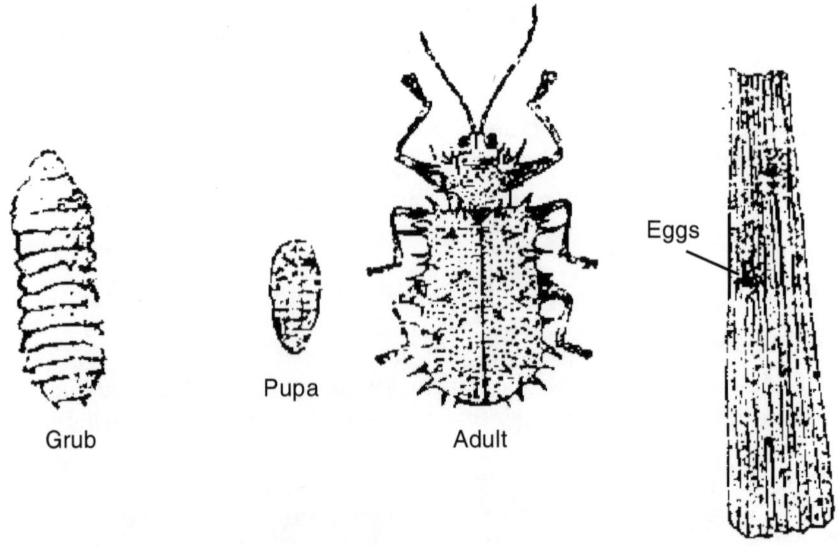

Figure 7.7: Rice Hispa

Nature of Damage

The grub mines into the leaf blade and feed on the green tissue between the veins. Adults also feed in the green tissue; they scrape the green matter of the tender leaves. Generally, the plants are affected in the young stage. The mining of the grubs will be clearly seen on the leaves. White parallel line will be clear on the leaves.

Life Cycle

The female starts egg lying in the nurseries. The eggs are embedded in the leaf tissue towards the tip. On hatching, the grubs feed as leaf-miners, between the upper and the lower epidermis. The attacked leaves are die. When the grubs are full-fed they pupate inside and finally emerge as black beetles. There are 2-6 generations in a year.

Management

1. The pest is suppressed if the infested leaf tips are clipped off and destroyed, while transplanting.
2. Spray chlorpyriphos 20 EC or fenitrothion 50 EC.

Wheat

Wheat Aphid, *Macrosiphum miscanthi* (Hemiptera: Aphididae)

Marks of Identification

The adults are green, inert, loucse like and first appear on the young leaves. The nymphs and the females look alike, except that the latter are larger. The winged forms appear only in early summer.

Nature of Damage

It attacks wheat, barley, and oats. The nymphs and adults suck the sap from plants, particularly from their ears and decrease yield of crop.

Life Cycle

The female give birth to young ones and are capable of reproducing without mating. During the active breeding season, there are no males and the rate of reproduction is very high.

Management

1. Excessive use of nitrogenous fertilizers should be avoided.
2. Wheat varieties *viz.*, HD-2329, Raj-3077, HD-2285, UP-2121, WH-542, HD-4645 are resistant to aphid.
3. The grubs and adults of ladybird beetle *Coccinella septumpunctata* feed voraciously on the nymphs and adults of aphid. Similarly the maggots of syrphid fly, feed very actively on aphid. Normally no control method for wheat aphid is required when these biocontrol agents are active.
4. Spray imidacloprid 17.8 per cent @ 3 ml per 10 litre of water.

Armyworm, *Mythimna separata* (Lepidoptera: Noctuidae)

Marks of Identification

The adult moths are pale brown in colour. The newly emerged larvae are dull white and later turn green.

Nature of Damage

The early stage larvae feed on tender leaves in the central whorl of the plant. As they grow they feed on older leaves of plants and skeletonize them totally. In case of severe attack whole leaves, including the mid-rib are consumed and the field looks as it grazed by cattle.

Life Cycle

The female lay eggs singly in rows or in clusters on dry or fresh plants, or on the soil. They hatch in 4-11 days depends upon weather. The larval period lasts for 13-14 days in spring and 88-100 days in winter. The insect spins a cocoon and pupate usually in soil. The pupal period lasts for 9-13 days. The adult live for 1-9 days.

Management

1. The pest can be suppressed by collecting and destroying the caterpillars.
2. Wheat varieties *viz.*, HD-1982, UP-2009, Sonalika, Kalyan Sona, UP-1008, UP-319, UP-215, UP-301, UP-368, WL-711 and C-306 have been found resistant to armyworm.
3. *Apantelis flavipes* is a larval parasitoid of the armyworm.
4. Spray 500 ml of dichlorovos 100 EC or one litre of quinalphos 25 EC in 425 litre of water to suppress this pest.

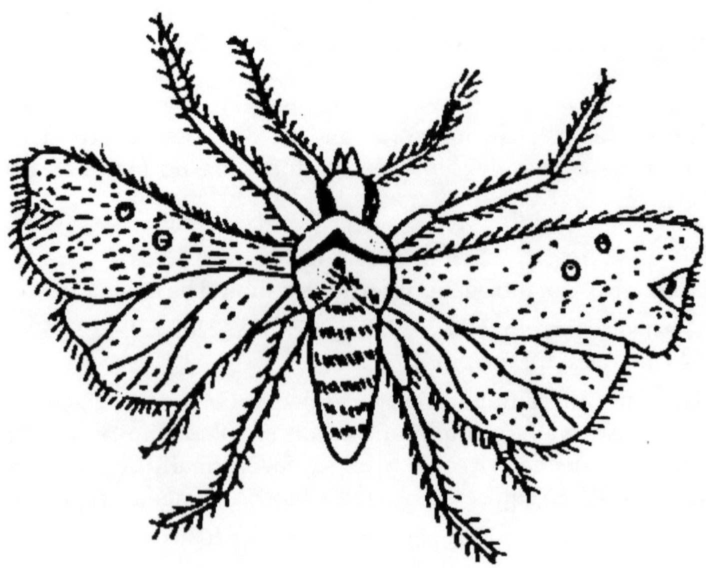

Figure 7.8: Armyworm

Ghujhia Weevil, *Tanymecus indicus* (Coleoptera: Curculionidae)

Marks of Identification

The weevils are earthen grey and measure about 6.8 mm in length and 2.4 mm in width. Their forewings are oblong and hindwings are more or less triangular. They can not fly.

Nature of Damage

The damage is caused by the adult weevils only and they cut the germinating seedlings at the ground levels. The damage is particularly serious during October-November.

Life Cycle

The female lays 6-76 eggs in 5-11 installment in the soil under clods in the ground. The eggs hatch in 6-7 weeks and young grubs enter the soil. The grubs feed on soil humus. They are full grown in 10-18 days and pupate in soil in earthen chamber. The pupal period lasts for 7-9 weeks. The pest complete only one generation in a year.

Management

1. Plough the field in summer to expose and kill the pupae.
2. Drenching of the soil should be done with chlorpyriphos 20 EC @ 4.5 litre per ha.
3. Mix thoroughly malathion 5 per cent dust @ 25 kg per ha with 12-25 cm deep layer of soil.

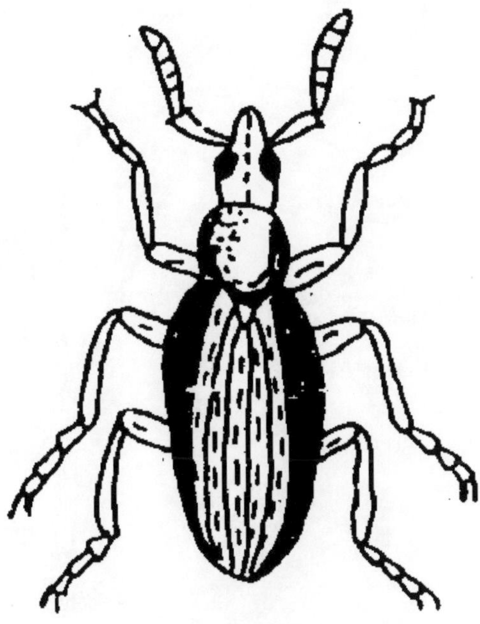

Figure 7.9: Ghujhia Weevil

Maize, Sorghum and Pearl Millet

Sorghum Shoot Fly, *Atherigona varia soccata* (Diptera : Anthomyiidae)

Marks of Identification

The adult is a small dark fly. Female fly has whitish grey head and thorax, while the abdomen is yellowish with paired brown patches. Male is darker in colour.

Nature of Damage

The maggots bore into the shoot of young plants, a week after germination to about one month and as a result the central shoot dries up. If the plants are attacked

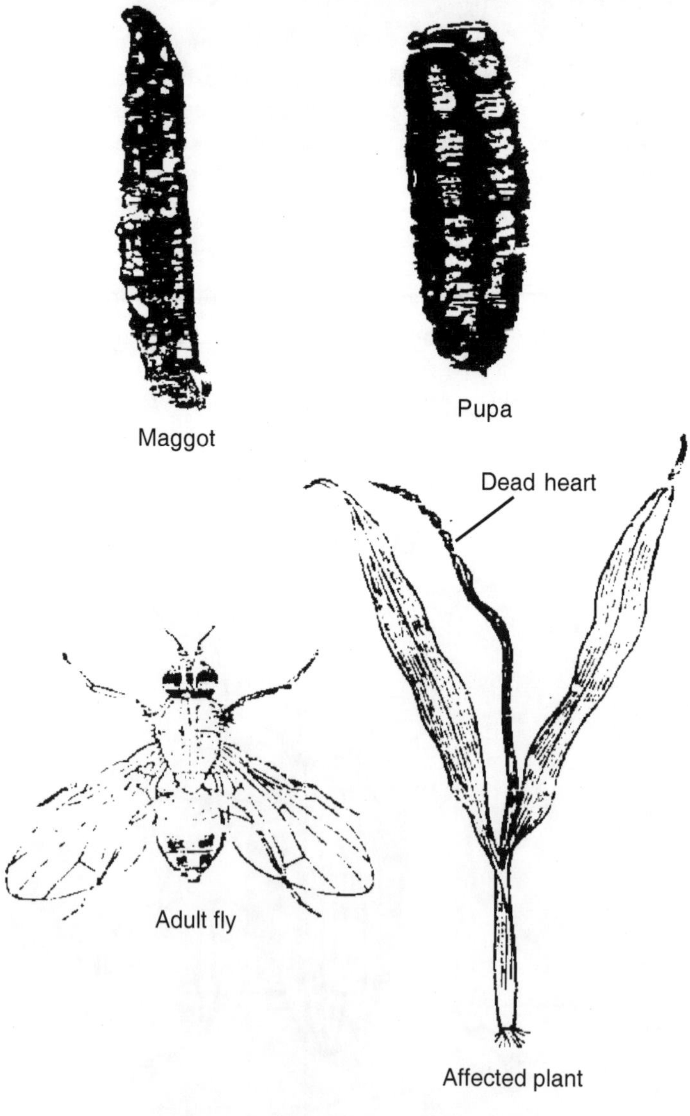

Figure 7.10: Sorghum Shoot Fly

at the initial stages the mother plant may produce profuse side tillers, but the tillers also may be attacked. The infestation often goes as high as 60 per cent. The high yielding hybrid varieties are severely attacked. In South India, crop is damaged during October to December as also in summer. Dead hearts or drying of central shoots or production of profuse side tillers in main plants is the main symptoms of its attack.

Life Cycle

The fly lays white minute (0.72 x 0.16 mm) rice grain shaped eggs sculptured with horizontal longitudinal ridges singly or in small batches on the stem above the ground or just below ground level and also on the under surface of cotyledonary or first leaf of young. The egg, larval and pupal stage of this pest lasted for 1-3, 15-18 and 8-9 days, respectively. Total life cycle was completed in 17-33 days. The adult flies survived for 1-2 weeks.

Management

1. Use higher seed rate to minimize the damage.
2. Later thinning out of infested and extra plants is also useful in areas which have low infestation.
3. Destruction of wild sorghum in the vicinity of field will reduce the natural breeding place of the pest.
4. Parasitoids *viz., Tetrastichus nyemitawus,* and *Neochryso-charis* spp. (Enlophidae) regulate pest population effectively under the field conditions.
5. A number of granular insecticides like carbofuran, fensulfothion, disulphoton, phorate, aldicarb, trichlorphon and carbaryl, in doses ranging as high as 1.0–2.5 kg a.i. per ha as soil furrow application at sowing time are effective.

Maize Stem Borer, *Chilo partellus* (Lepidoptera: Pyralidae)

Marks of Identification

Moth is medium sized and straw coloured. Male has pale brown forewings provided with dark brown scales forming a dark area along the coastal margin. Hindwings are light straw in colour. Female possesses forewing of a lighter colour and nearly white hind wings.

Nature of Damage

The caterpillar bores into the stem and feeds on the central shoot. There may be more than one caterpillar in a single plant. In early stages, the caterpillars make circular holes on unfolded leaves and later central shoot dries up producing dead heart. Later it acts as an internode borer and is found till the time of harvest. Young cobs may also be attacked. Yield is affected much and the quality of the fodder is also reduced. The damage caused to the crop by this pest was estimated to range between 70–80 per cent. Presence of circular holes on the unfolded leaves and dead hearts in the early stages are the main symptoms. The boreholes may be visible in contrast to the dead heart caused by the stem borer. When grown up plants are attacked the symptoms will not be quite visible.

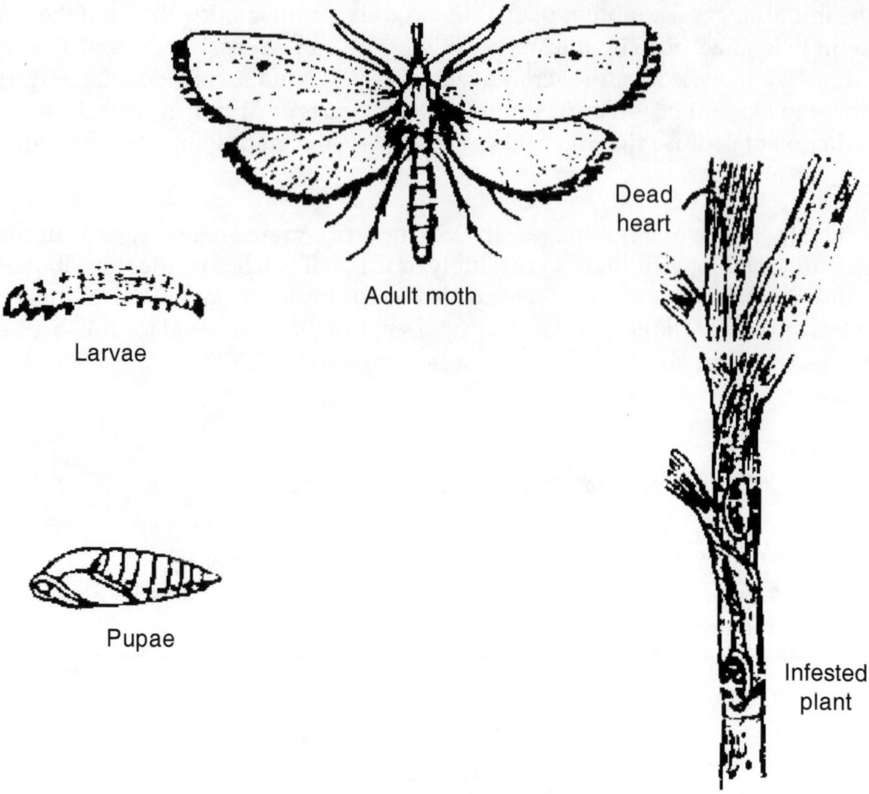

Figure 7.11: Maize Stem Borer

Life Cycle

The average fecundity per female is 87 to 110 eggs. Egg are laid in masses of 20-25 on the underside of the leaves. The egg, larva, pupal stage lasts for 4-5, 18-20, 6-8 days, respectively, and life cycle is completed in 29-33 days in case of non-hibernating stage during summer while corresponding figures are 8-9, 28-93, 9-24, 83-210 days, respectively. There are 6 larval instars. The full grown caterpillars of the last generation hibernate in stubble, stalks etc. and remain there till the next season. There are probably 5 generations in a year.

Management

1. The manipulation of the date of sowing, higher plant population and use of sorghum as a trap crop can help in achieving the targeted yield of maize.
2. Plough up the field soon after harvesting the maize, *Jowar* and *Bajra* crop. Collect and burn the stubbles. Use the stalks by the end of February and chop the remaining stalks for subsequent use. The cores of maize cobs may harbor the borer larvae. These should also be destroyed by burning in the end of February. For seed, keep only healthy cobs free from borer.

3. Clipping of lower leaves of maize (up to fourth) on which most of eggs of *C. partellus* are laid reduces damage to some extent.
4. Use the recommended seed rate.
5. Remove and destroy the plants showing severe borer injury while hoeing the crop.
6. Growing maize in association with various legumes significantly reduce pest incidence in maize. Also inter-cropping maize with soybean give considerable reduction in pest attack.
7. Spray the crop 2-3 weeks after sowing or as soon as borer injury to the leaves is noticed with fenvalerate 20 EC, cypermethrin10 EC, or deltamethrin 2.8 EC.

Pink Stem Borer, *Sesamia inferens* (Lepidoptera : Noctuidae)

Marks of Identification

Adults are stout, straw coloured and are nocturnal in habit. The fully developed caterpillar is cylindrical, pinkish dorsally and whitish ventrally. Larvae can migrate from plant to plant.

Nature of Damage

The young larvae after hatching, congregate inside the leaf whorls and feed on folded central leaves causing typical 'pin hole' symptoms. Severe feeding results in killing of the central shoot and consequent dead heart formation. Usually the second instar larvae migrate to neighboring plants by coming out from the whorls and suspending themselves from the plants by silken threads, these are then easily blown off by wind to other plants. These larvae penetrate in the stem and cause tunneling resulting in stunting, infested plants become weak and bear very small earheads. The weakened stems, especially of tall local varieties, break easily during heavy rains or with high velocity winds.

Life Cycle

The female lays *bead* like eggs in 2 to 3 longitudinal rows within the lower most leaf sheaths preferably of young plants of maize. The incubation period is 6 to 7 days. The larval period varies from 24 to 36 days. The pupal period is about 10 days. The total life cycle varies from 40 to 80 days. The larval period is prolonged in severe winter. The adults live for 4-6 days

Management

1. Removal and destruction of dead hearted plant is recommended.
2. Sex pheromone traps with (Z)-11-hexadecenyl acetate and (Z)-11-Hexadecen-1-0 L in a ratio of 4:1 is effective method for forecasting the occurrence of *S.inferens.*
3. Maize varieties 'Syn A 226' Syn E 13, 'Puerto Rico Gr 1 x TAD', Puerto Rico Gr 1 x E 13; Ant. 40; Kitole III; H 2; JML 22; Syn B 19; Thai DMR; Warangal

local; TAD x Syn 13; DHM 103; EH 4025; Mex 17; Hybrid Odesski 50; Temp. Yellow dent; Ganga 5; Antigua Gr.1 etc. are fairly resistant.

4. Pink stem borer is parasitized by *Telenomus* sp. and *Trichogramma australicum* in egg stage, *Apanteles flavipes, Bracon chinensis* in larval stage and *Tetrastichus ayyari* and *Xanthopimpla* sp. in pupal stage.
5. Spray the crop 2-3 weeks after sowing or as soon as borer injury to the leaves is noticed with fenvalerate 20 EC, cypermethrin10 EC, or deltamethrin 2.8 EC.

Sorghum Earhead Bug, *Calocoris angustatus* (Hemiptera : Miridae)

Marks of Identification

The adult is a slender green elongate bug about 1 cm long and active flier. The newly hatched nymphs have light orange red abdomen, which changes to green in the advanced instars.

Nature of Damage

The adults and nymphs live inside the earhead and suck the milky fluid from the tender ripening grains. Due to the feeding, the grains get shriveled and chaffy and thus unfit for sowing and for consumption. No damage is caused to fully ripened grains. A reduction of 15–30 per cent in the yield was estimated due to its attack. Usually high yielding varieties with compact earheads are subjected to more infestation than the loose earheads. No external symptom will be visible. The earheads should be tapped either on the palm or a piece of cardboard. A number of brownish or greenish nymphs and adults can be seen. On the developing grains small brownish spots will be visible. In severe infestation, the grains get shriveled without maturing and the earheads appear uneven.

Life Cycle

The adults appear on sorghum crop as soon as the ears emerge from the leaf sheaths. The bugs lay eggs under the glumes or in between anthers of florets, by inserting its ovipositor. The female lays 150-200 cigar shape eggs. The eggs hatch in 5-7 days and the nymphs start feeding on developing grains in the milk stage. The nymphs pass through 5 instars and develop into adults in about 3 weeks. The insect completes its life cycle in about one month and there are many generations in a year.

Management

1. Avoid staggered sowing in an area which favour multiplication of bug.
2. Setup light trap till mid-night to attract and kill adults.
3. Apply malathion 5 per cent or phosalone 4 per cent dust (@ 25 kg per ha 3 and 18 days after panicles emergence.

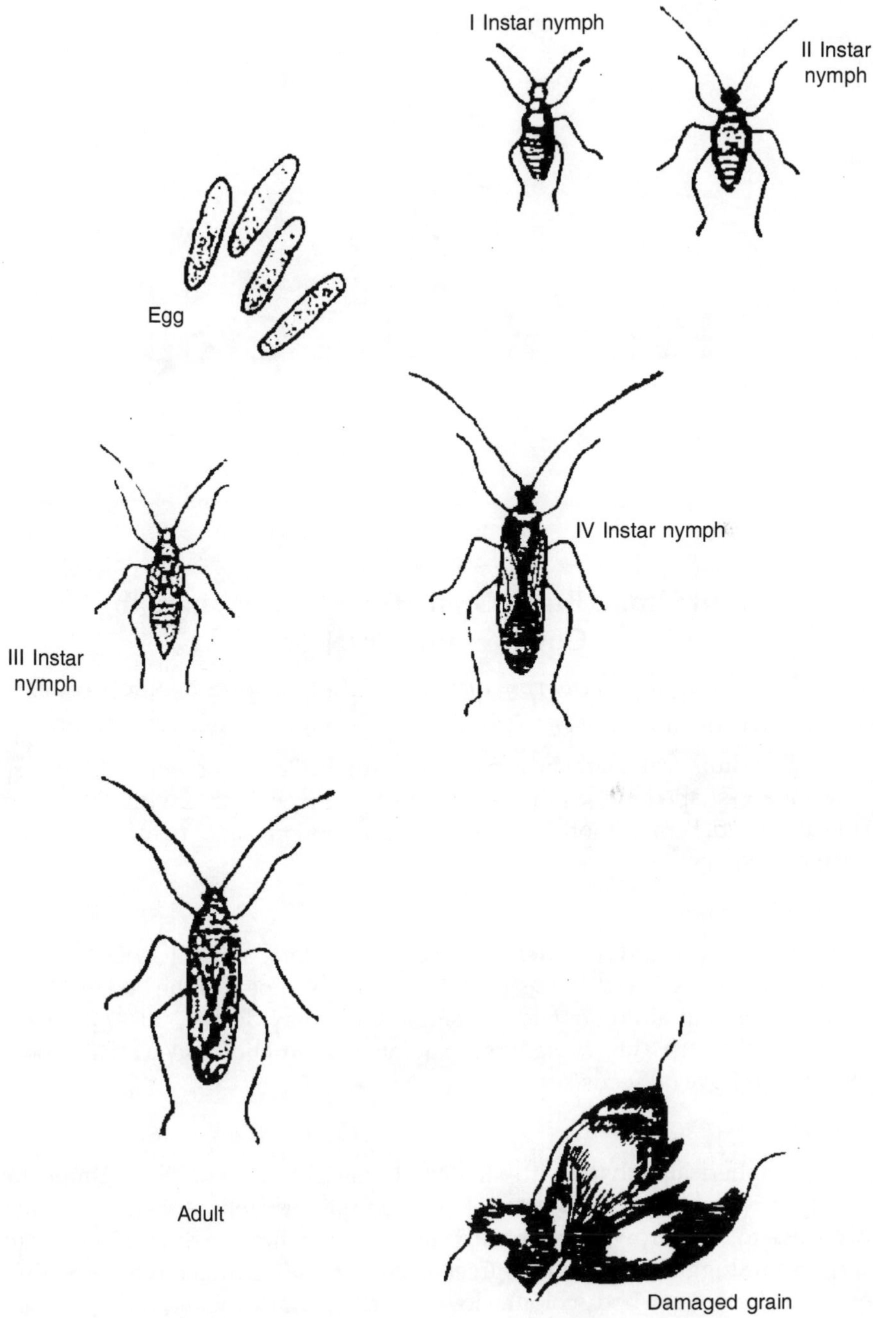

Figure 7.12: Sorghum Earhead Bug

8

Pest of Pulse Crop

Red Gram, Black Gram, Green Gram, Lablab, Cowpea and Chickpea

Gram Pod Borer, *Helicoverpa armigera*, (Lepidoptera: Noctuidae)

Marks of Identification

It is a medium-sized light brown coloured moth. On the forewings, there is speck that forms a V-shaped mark. Hind wings are dull grey coloured with a black border on the distal end. Female moth is bigger than male and presence of tuft of hairs on the tip of the abdomen.

Nature of Damage

Young larva feeds on tender leaves, buds, flowers, and subsequently it bores into the pods and feeds on the seeds with its head and part of the body only thrust inside, the rest remaining outside. A single larva may destroy 30-40 pods before maturity. In the early stages, plants seen defoliated. Boreholes seen on the pods and affected pods have no seeds.

Life Cycle

Egg is spherical in shape with a flattened base, giving dome shaped appearance, and surface is sculptured in the form of longitudinal ribs. Yellowish-white, glistening and change to dark brown, before hatching. They hatch in 2-6 days. Newly hatched caterpillar is sluggish and whitish-green in colour. Full-grown larva is 3.5-4.0 cm in length with pale-green body colour. However, the colour varies according to the food intake. Dorsal surface bears dark brown stripes. Head is reddish-brown. Larva is highly cannibalistic and readily eats one another. The larvae full fed in 13-19 days. It pupates in soil in earthen cell. Pupa is obtect type. Freshly formed pupa is greenish

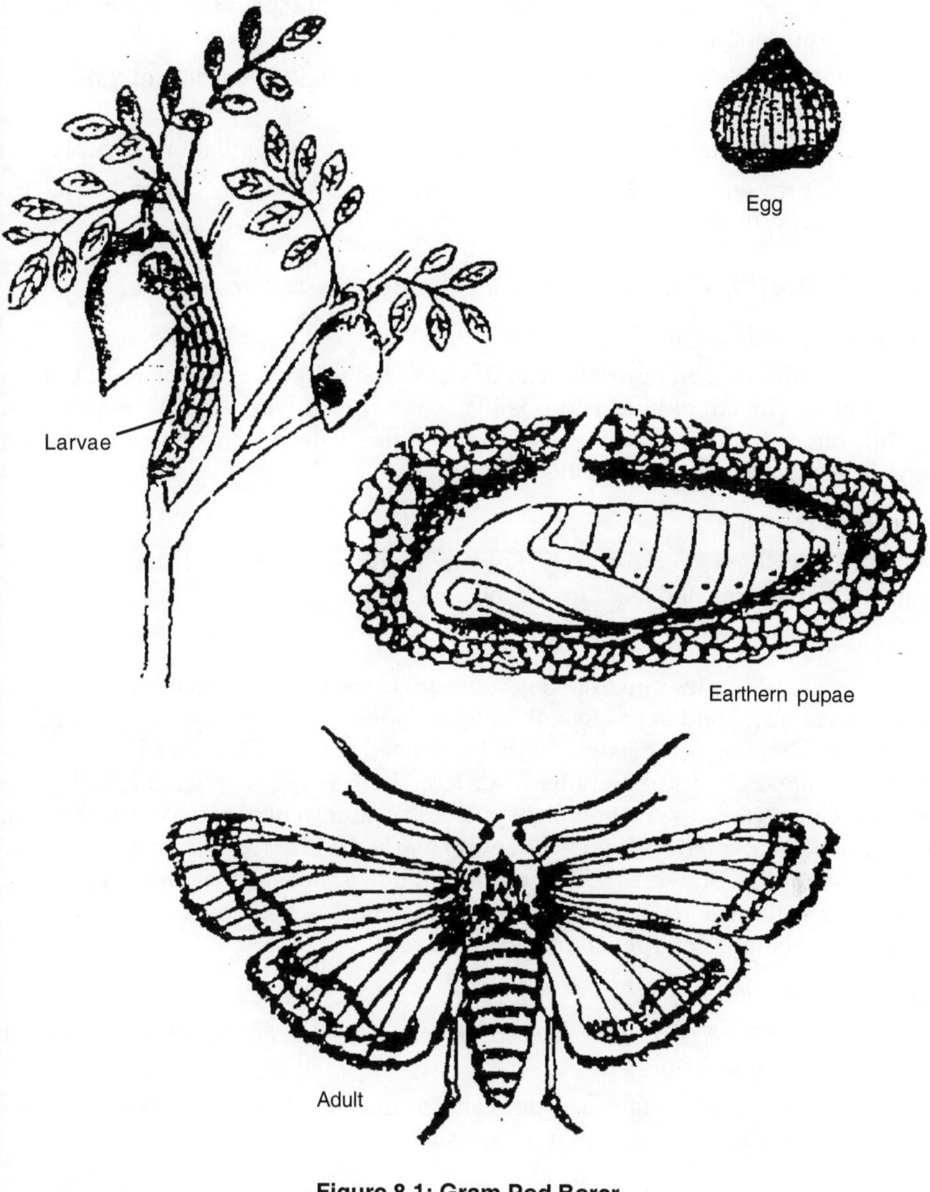

Figure 8.1: Gram Pod Borer

yellow in colour and darkened prior to emergence of moths. The pupal period lasts for 8-15 days. There mat be as many as 8 generations in a year.

Management

1. Timely sowing *i.e.,* upto mid October or growing early maturing varieties which complete podding by first week of March.

2. Mixed or inter cropping with non- preferred host plants like barley, wheat, mustard and linseed.
3. Apply *Helicoverpa* NPV @ 250-500 LE per ha alone or half dose of endosulfan 35 EC (1.25 litre per ha).
4. Install pheromone traps for monitoring the pest population (12 per ha).
5. Spray endosulfan 35 EC (2.5 litre per ha) or fenvalerate (250 ml per ha) or cypermethrin (200 ml per ha) etc.

Blue Butterfly, *Lampides boeticus* (Lepidoptera: Lycaenidae)

Marks of Identification

It is medium sized butterfly. The colour of the wings is violet metallic blue to dusky blue. The tail of hind wings is black and tipped with white. The female is slightly bigger than the male. In males, the abdomen is slender and tapering, while in female it is long and broader at the tip.

Nature of Damage

The larva bores into the buds, flowers and green pods just within couple of hours after hatching and feeds inside the developing grains.

Life Cycle

Eggs are laid on the buds, flowers, green pods and on shoot and leaves. Greenish white in colour, round in shape with a slight depression at the top. Newly hatched larva is yellowish green in colour with black head and a dark-brown patch on the prothorax and cylindrical body with scattered hair. Full-grown larva is yellowish green to yellowish red sometimes light purple in colour, ventral surface is light green. Whole larva is covered with small setae and marked with irregular black markings. It looks like a slug. Brownish mid-dorsal and yellowish lateral lines are well marked. Pupa are green in colour later on it darkens and wings are also visible. Five weeks are required to complete one generation in field conditions.

Management

1. Spraying 5 per cent neem seed kernel suspension give good results against this butterfly pest.
2. Spray endosulfan (0.07 per cent) to manage this pest in case of severe infestation.

Plume Moth, *Exelastis atomosa* (Lepidoptera: Pterophoridae)

Marks of Identification

It is lightly built and light brown in colour, wings deeply fissured, the forewings longitudinally cleft into two plumes and hind wings into three plumes. Forewings are extremely elongate. Legs are long and slender. Abdomen is dark-brown in colour.

Nature of Damage

Pods are scrapped in the early stages, later boreholes seen on the pods and seeds

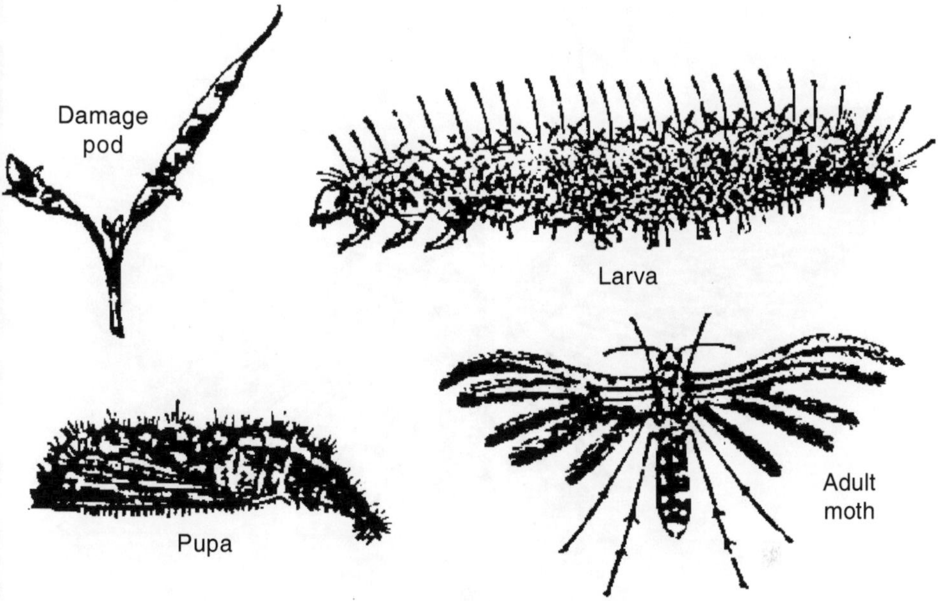

Figure 8.2: Plume Moth

eaten away. Young larvae bore into the unopened flower buds for consuming the developing anthers. Grown up larvae first scrap the surface of the pods and then bore into pods. The larvae never enter the pod completely.

Life Cycle

The female moths lay 17-19 eggs singly on tender parts of the plants. The eggs hatch in 2-5 days and the young larvae feed on the pods and become full grown in 10-26 days. Pupation takes place outside the pod on its surface or in the entrance hole ifself. The pupal period lasts for 3-12 days. The life cycle is completed in 17-42 days.

Management

1. Spray endosulfan 35 EC (2.5 litre per ha) or fenvalerate (250 ml per ha) or cypermethrin (200 ml per ha).

Spotted Pod Borer, *Maruca testulalis* (Lepidoptera: Pyraustidae)

Marks of Identification

Dark brown with a white cross band in the middle of the forewings and the hind wings are white with a darker border.

Nature of Damage

Presence of semi-solid excreta at the junction of the borehole. Young shoot with dried tip, large scale dropping of flowers. Larva present inside the webbing of leaves, flowers and young pods, faecal material accumulates outside the borehole. It feeds on the seeds by boring into the pods.

Figure 8.3: Spotted Pod Borer

Life Cycle

Eggs are elongate oval in shape and light yellow in colour and are laid singly on or near flower buds of host plants. Young caterpillars feed on reproductive parts of flowers and move from one flower to another. Later they web inflorescences with adjacent leaves and developing pods and feed within by boring into the flowers and pods. Full grown caterpillars are light brown in colour with irregular brownish black dorsal, lateral and ventral spots. Incubation period is 2-3 days. Larval stage lasts for 8-14 days and pupal period lasts for 6-9 days.

Management

1. Spray endosulfan 35 EC (2.5 litre per ha) or fenvalerate (250 ml per ha) or cypermethrin (200 ml per ha).

Spiny Pod Borer, *Etiella zinckenella* (Lepidoptera: Phycitidae)

Marks of Identifications

Greyish brown moth, distinct pale-white band along the costal margin of the forewings, hind wings are semi-transparent with a dark marginal line. Prothorax is orange in colour.

Nature of Damage

Entrance hole in the green pod disappears and leaves little evidence that the pod is infested. In pods, the larva devours many seeds. The pod always contains a mass of frass and held together by a loosely spun web. Young larva bores into floral parts, making rough and irregular incision.

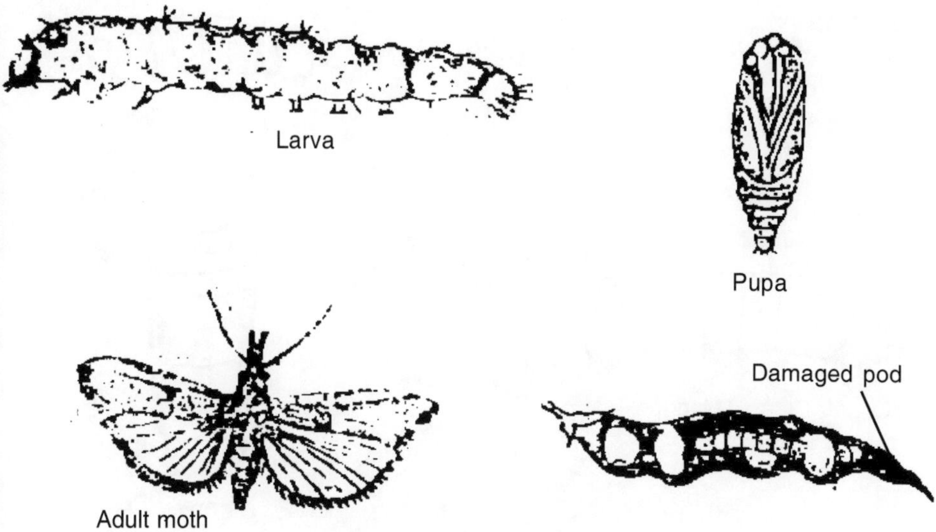

Figure 8.4: Spiny Pod Borer

Life Cycle

A female can lay 50-200 eggs. The eggs are laid singly or in clusters on any part of the plant including pods. Incubation period lasts for 5-6 days. The newly hatched larvae are greenish first feed on flowers and then bore into the pods to feed the seeds. The larvae full-fed in 10-13 days. Pupation occurs at a depth of 2-4 cm under the soil. Pupal period lasts for 9-20 days. Overwintering pupae emerge in February or March. Adults live for 5-7 days.

Management

1. Spraying 5 per cent neem kernel suspension give good results against this pest.
2. Spray endosulfan (0.07 per cent) to manage this pest in case of severe infestation.

Tur Podfly, *Melanagromyza obtuse* (Diptera: Agromyzidae)

Marks of Identifications

It is slightly bigger than the male. Its wings are also slightly broader. Colour of a newly emerged adult is dull-white and smoky patches at places, but gradually it acquires the normal black colour with slight bright greenish tinge. Abdomen is glossy black, but in some cases it is slight bronzy, while in other cases it has a greenish-blue background.

Nature of Damage

Shriveled pods and seeds. Damaged seeds become unfit for consumption and also do not germinate. However, the attack of the fly remains unnoticed by the farmers

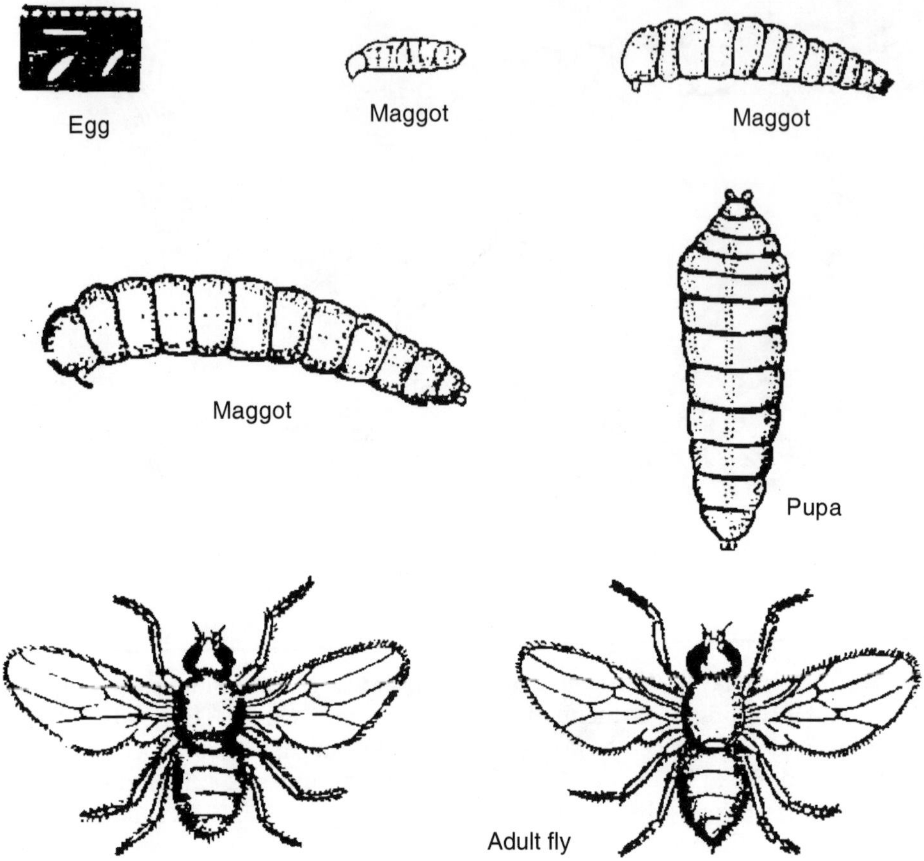

Figure 8.5: Tur Podfly

due to the concealed mode of life of this insect within the pods. Young maggot attaches itself on the immature seed inside the pod. In the beginning it feeds on the surface and thereafter mines into seeds and makes galleries just under the seeds epidermis, causing a ring like track. (One seed is enough for the development of a maggot. It never leaves the pod in which it enters once and completes its maggot stage).

Life Cycle

Freshly laid eggs are white, broad and round at its posterior end which is embedded in the tissues of the pod and narrowed anteriorly into a somewhat elongated egg sheath. Eggs hatch in 3-5 days. Freshly hatched maggot is white with dark-brown mouth hook. Full-grown maggots is cylindrical in shape and is narrower at the head end, which bears black mouthparts. It is creamy white but acquires a yellowish tinge just before pupation. Maggots become full-fed in 8-12 days. Pupation takes place in hard chitinous puparium, which is found sticking to the side of pod or in the groove eaten into the grain by the larva. It is cylindrical with broadly rounded ends. Fresh pupa is yellowish white, but becomes darker subsequently. Pupal period lasts for 7-10 days. The life cycle completed in 11-27 days and several generations are produced in a year.

Management

1. Spray of endosulfan (0.07 per cent) followed by monocrotophos (0.04 per cent) at some intervals during flowering and pod-formation.

Stem Fly, *Ophiomyia phaseoli* (Diptera: Agromyzidae)

Marks of Identification

Light brown when freshly emerged, but fully developed adult is metallic-bluish or greenish-black in colour with light brown eyes. Wings are transparent. Female is slightly bigger than the male.

Figure 8.6: Stem Fly

Nature of Damage

Drooping of the tender leaves and yellowing is the serious damage of young plants. The sites where maggot and pupae are present become swollen and start rotting. Older plants show stunting but are not usually killed. Maggot is the damaging stage. It mines sub-epidermally through the leaves. Plants are most seriously affected at the seedling stage, where stem is tunneled.

Life Cycle

The female lay eggs in holes made on the upper surface of young leaves, especially near the petiol end of the leaf. On hatching the maggot forms a short linear leaf mine and further on it tunnels underneath the epidermis of the leaf until it reaches one of the veins which leads it to the midrib and then to the leaf stalk and the stem. Pupation takes place inside the stem. The total life cycle take 2–3 weeks. As many as 7 generations have been completed in a year.

Management

1. Spray monocrotophos 0.04 per cent during the flowering stage.
2. Soil application of 10 kg of phorate 10G is effective upto 40 days of sowing.

Blister Beetle, *Mylabris pustulata* (Coleoptera: Meloidae)

Marks of Identification

Medium sized, 12.5-25.0 mm long, conspicuous in appearance and are moderately robustly built. Beetles are bright metallic blue, green, black and yellow or brown in colour.

Nature of Damage

The adult beetles feed on flowers, leaves and tender panicles, thus preventing grain formation.

Life Cycle

Eggs laid on the ground or in the soil. First stage larva is 'triungulins' (long-legged) and actively searches for the host. They moult to become cruciform or caraboid (Hyper metamorphosis). Pupation takes place in soil.

Management

1. Manual picking and destruction of adult bleaster beetles is often the only practical control measure.
2. In case of severe infestation spray monocrotophos 0.04 per cent gives good results.

Bean Aphid, *Aphis craccivora*, (Hemiptera: Aphididae)

Marks of Identification

Apterous females are shiny, dark brown or black. Alate forms are greenish black with transparent wings.

Nature of Damage

Colonies of nymphs and adults found on leaves, terminal shoots and pods and suck the plant sap. They also act as vector of stunt disease in chickpea, rosette of groundnut. Under low rainfall situations it becomes a serious pest.

Life Cycle

The common mode of reproduction is through vivipary and parthenogenesis, though reproduction by ovipary has been also recorded. A single apterous, parthenogenetic female produced 29 nymphs. The nymphs generally undergo 4 moults before reaching the adult stage. The duration of each instar is usually one day, though in some cases it was even 3 days. Within a day after it becomes an adult the apterous female starts producing its brood. A female reproduced upto a maximum of 12 days.

Management

1. Encourage the activities of the predators, *viz.*, *C. septunpunctata*, *M. sexmaculatus*, *B. suturalis* and *X. scutellarae.*
2. Spray metasystox (0.1 per cent) or demeton-o-methyl (0.05 per cent).

Pod Bug, *Clavigralla gibbosa*, (Hemiptera: Coreidae)

Marks of Identifications

The adult bugs are greenish-brown in colour, 2 cm in length, with spines on either side of the middle of the prothorax. Female bug is bigger and has a round and swollen abdomen in comparison with a narrow and pointed abdomen of the male.

Nature of Damage

Nymphs and adults cause substantial damage to pods and also to stem, leaves and flower buds. Attacked pods show pale-yellow patches. When the attack is heavy, the pods shrivel up. The grains in the attacked pods remain shriveled and extremely small. Both the nymphs and adults cause damage by sucking juice. The pest assumes serious proportions on the pods before the maturity of the crop.

Life Cycle

The eggs are preferably laid on pods. They hatch in 7-8 days. On hatching the nymphs collect at a point on any part of the plant to feed gregariously including pods. The nymphal development completes in 17-20 days. Adult bug live for 6-7 days. The pest is active in the months of October-May.

Management

1. The eggs of this pest are parasitized by *Gryon* (= *Hydronotus*) *antestiae* (Hymenoptera: Scelionidae).
2. Spray the crop with monocrotophos (0.04 per cent).

9
Pest of Oilseed

Mustard

Mustard Aphid, *Lipaphis erysimi* (Hemiptera: Aphididae)

Marks of Identification

Minute, soft bodied, light green and pear shaped insects having a pair of short tubes called cornicles or honey tubes or siphons on the postero-dorsal region of the abdomen.

Nature of Damage

Both nymphs and adults damage the plants by sucking sap from leaves, buds and pods. Due to very high pest population, the vitality of plants is greatly reduced, leaves become pale and curled up, flowers fall to form pods and developing pods fail to form seeds so that the yield may reduce upto 25-40 per cent.

Life Cycle

The main season of activity of the aphid coincides with the general growing period of cruciferous crops, extending from October to March. During this period it completes on an average 11 generations. The pre reproduction and reproduction period, 8-18 and 5-7 days, respectively. The total life cycle completes in 23-67 days.

Management

1. Early sowing of mustard is preferred.
2. Under field conditions *D. rapae* is very effective parasitoid of this aphid. Predators *viz.*, *C. septunpunctata*, *M. sexmaculatus*, *B. suturalis* and *X. scutellarae* check the population of this pest effectively in field. So there activities should be encouraged.

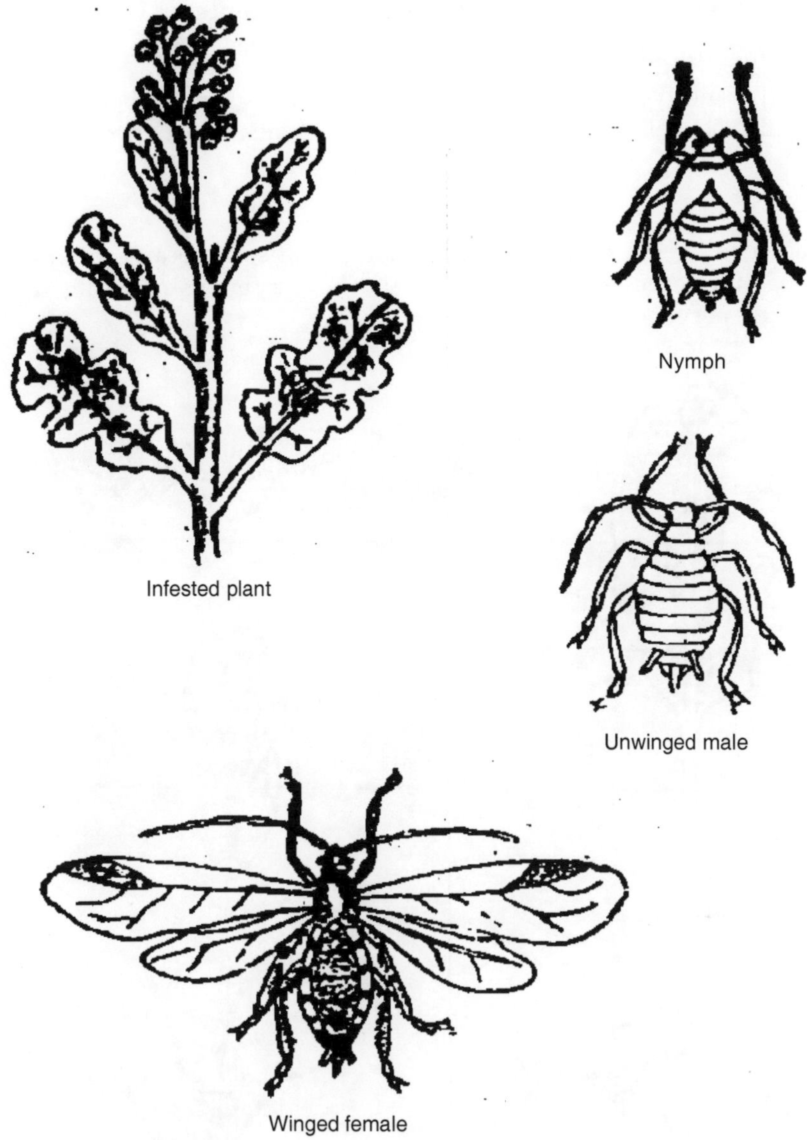

Figure 9.1: Mustard Aphid

3. Mustard cultivars *viz.*, T-59, B-85 glossy, RW white glossy and RL-18, RLM-198 are resistant to this pest.
4. Spray neem seed *kernel* extract (5 per cent) or neem oil (0.5 and 1.0 per cent).
5. Spray dimethoate (0.03 per cent) or endosulfan (0.05 per cent) or methyl demeton (0.025 per cent).

Mustard Sawfly, *Athalia lugens proxima* (Hymenoptera : Tenthredinidae)

Marks of Identification

Small orange yellowish insects with black markings on the body and smoky wings with black veins. Larvae are dark-green with black dorsal strips and wrinkled body.

Nature of Damage

Rapeseed and mustard seedlings are seriously affected because the grub either riddle the entire leaves or cause numerous shot holes by excessive feeding. This pest causes 12-13 per cent loss in the yield.

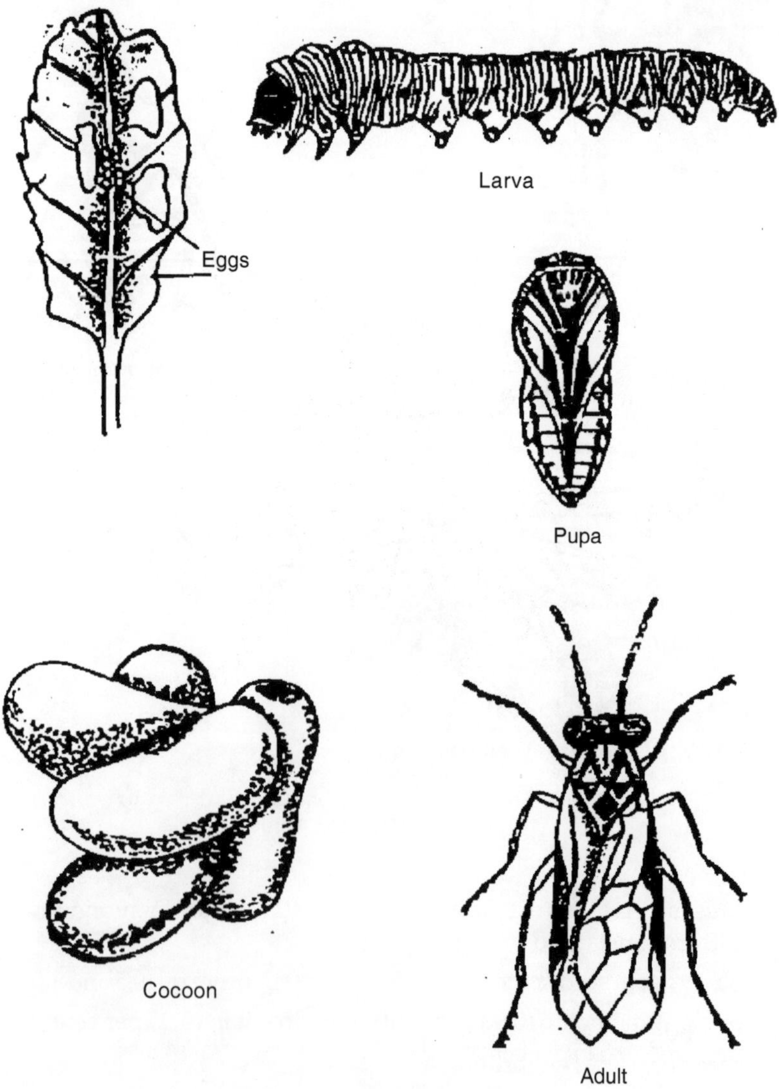

Figure 9.2: Mustard Sawfly

Life Cycle

About a week after the first mating the female start laying eggs, which are inserted singly in the tissues of leaves around the periphery, within incisions with the help of the ovipositor. The incubation period varied from 3-6 days. The larval period lasts for 7-19 days. One female lay about 40 eggs. There are six larval instars. The full grown grubs enter the soil and form elongated oval cocoons. The larvae prefer sandy soil with 15 per cent moisture for pupation. Pupal period lasts for 7-10 days. The adults live upto a maximum of 20 days. The pest complete 3-4 generations in a year.

Management

1. Saw fly infestation could be minimized by sowing mustard crop late (5th November).
2. Pest incidence could be minimized by intercropping mustard with gram.
3. The activities of the grub could be reduced by irrigating the crop after 15 days of sowing.
4. The nymphs and the adults of *Cantheconida furcellata* attack the grubs. Larval pupal parasitoid, *Peritissus cingulator* (Hymenoptera: Ichneumoidae) parasitize the larvae of the pest. The bacterium, *Serratia marcescens* caused mortality of the grubs.
5. Spray neem seed kernel extract (5 per cent) or Mahua (*Madhuca indica*) suppress the pest population effectively.
6. Spray insecticides *viz.*, fenvalerate (0.005 per cent) or monocrotophos (0.04 per cent) for effective control of this pest.

Painted Bug, *Bagrada cruciferarum* (Hemiptera : Pentatomidae)

Marks of Identifications

The full-grown nymphs are about 4 mm long and 2.66 mm broad. They are black with a number of brown markings. The adult bugs are 3.71 mm long and 3.33 mm broad. They are sub-ovate, black and have a number of orange or brownish spots.

Nature of Damage

Both nymphs and adults of this bug suck sap, especially from cruciferous plants and devitalize them, when attacked young plants wilt and die. On larger plants, bugs cluster over the leaves and pods and suck the sap. These bugs also feed on mustard seeds.

Life Cycle

A female on an average lay 32-100 eggs in her life span of 2-4 weeks. The egg stage is 3-20 days. The nymphal period is 14-36 days and the nymphs passed through 5–6 instars. There are 9 generations in a year.

Management

1. The bug is parasitized by a techinid, *Alophora* sp. while the eggs are parasitized by Scelionids, *Gryon* sp., *Trissoleus samueli* and *Typhodytes* sp.

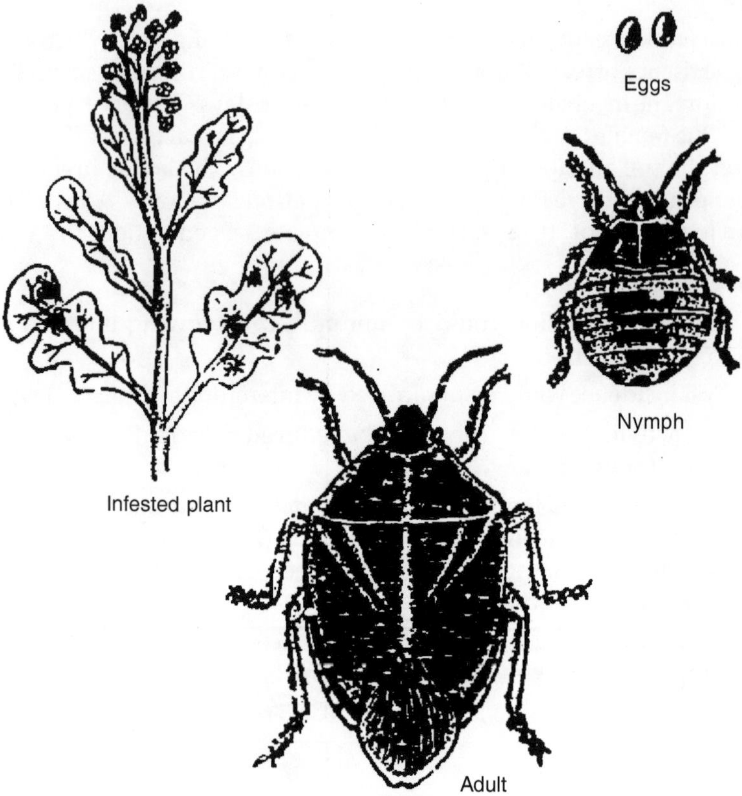

Figure 9.3: Painted Bug

2. Spray *Heeng* (*Asafoetida*) @ 20 g per litre or neem based pesticides effectiviely control this pest.
3. Dust endosulfan (4 per cent) or methyl parathion (2 per cent) @ 20 kg per ha at the time of seedling stage control the pest effectively.

Groundnut

Aphid, *Aphis craccivora* (Hemiptera: Aphididae)

Marks of Identification

Apterous females are shiny, dark brown or black in colour. Alate forms are greenish black with transparent wings.

Nature of Damage

The adults and nymphs feed on the growing tips before the crop starts flowering. Later they migrate to the floral shoots, seriously hampering pod formation. It is estimated that this pest alone reduces the yield by about 40 per cent. It is also a vector of *rosette*, a viral disease of groundnut

Life Cycle

The common mode of reproduction is through vivipary and parthenogenesis, though reproduction by ovipary has been also recorded. A single apterous, parthenogenetic female produced 29 nymphs. The nymphs generally undergo 4 moults before reaching the adult stage. The duration of each instar is usually one day, though in some cases it was even 3 days. Within a day after it becomes an adult the apterous female starts producing its brood. A female reproduced upto a maximum of 12 days.

Management

1. Groundnut intercropped with cereals (*e.g.* millets or maize)reduced the incidence of this pest.
2. Encourage the activities of the predators, *viz.*, *C. septumpunctata*, *M. sexmaculatus*, *B. suturalis* and *X. scutellarae.*
3. Spray metasystox (0.1 per cent) or demeton-o-methyl (0.05 per cent).

White Grub, *Holotrichia* spp. (Coleoptera: Scarabaeidae)

Marks of Identification

The full grown grubs are about 35 mm long and white, having brown head and prominent thoracic legs. The adult beetles are dull brown and measure about 18 mm in length and 7 mm in width.

Nature of Damage

The grub eat away the nodules, the fine rootlets and may also girdle the main root, ultimately killing the plant. The beetles also defoliate the plants *viz.*, neem, banyan, etc.

Life Cycle

The adult beetle lay eggs singly in the soil upto 10 cm deep. The incubation period ranged from 7-10 days. Eggs are laid with the onset of monsoon in July. There are three grub instars. The full grown grubs go deep into the soil to pupate. The pupal period takes 13 days. There is only one generation in a year. The beetle hibernates in the soil and come out of it with pre-monsoon shower in June. The beetle emerge from the soil between 7.30 to 8.00 PM and immediately fly to their host plants.

Management

1. Plough the field twice during May-June. It would help in exposing the beetles resting in the soil.
2. Wherever possible, sow the crop early *i.e.*, between June 10 to 20.
3. Treat the seed before sowing with 12.5 ml of chlorpyriphos 20 EC per kg of kernels.
4. Per-sowing soil treatment with phorate 10G at 25 kg per ha or quinalphos 5G at 30 kg per ha effectively protect the crop from the attack of grub.

Castor

Castor Semilooper, *Achaea janata* (Lepidoptera: Noctuidae)

Marks of Identification

The moths are stoutly build with smoky grey or brown forewings. Hindwings are dark with a white band in the middle and three to four white spots at the anal margin. The larva is a semilooper. Body is grey or black in colour, with red or whitish side strips. A full grown larva measures about 60-70 mm in length. The pupa is ashy grey. The colour pattern of third and fourth instar larvae is variable.

Adult moth

Adult moth

Larva

Figure 9.4: Castor Semilooper

Nature of Damage

The first instar larvae nibble the outer tissues of the leaves while the second stage larvae bite holes on castor leaves. The later instar larvae eat the leaves completely leaving behind only veins and petioles. Normally, it does not attack stem and branches, however, under severe infestation, the larva damages even inflorescence and young capsules. The moths puncture citrus fruits at night and suck the juice.

Life Cycle

The female moth lays about 450 eggs singly and scattered over the lower surface of leaves. The eggs are round, concave from the under side, convex from below and pale green in colour. The incubation period is 3-4 days. The newly hatched larva starts feeding on leaves and full grown in 15-20 days after 4-5 moultings. Pupation takes place in dried leaves either on plant or fallen in soil in a cocoon. The pupal period lasts 10-25 days, the longevity of adults range between 7-19 days. There are 5-6 generations in a year. The infestation may starts from July and continues till December.

Management

1. Light ploughing after harvesting of castor crop helps in reducing the pupal population.
2. The crop should be seeded during first fortnight of August so that the crop may escape from the incidence of this pest.
3. Light trap is one of the best tool to attract and destroy the adult moths.
4. Hand collection of big sized larvae helps in large scale reduction of the pest.
5. Egg parasitoids *viz., Trichogramma chilonis, T. achae, Telenomus* spp. and *Trissolcus* are the most effective eggs parasitoids of semilooper egg. Therefore, Inundative release of *Trichogramma* @ 100000 per ha per week should be made after the onset of moth emergence. *Microplitis maculipennis* is also effective larval parasitoid.
6. The crop should be sprayed with monocrotophos 36 WS 0.04 per cent or quinalphos 25 EC 0.05 per cent or endosulfan 35 EC 0.07 per cent based on ETL (4 larvae per plant).

Shoot and Capsule Borer, *Conogethes (=Dichocrocis) punctiferalis*, (Lepidoptera: Pyraustidae)

Marks of Identification

The adult is a small sized bright-yellow coloured moth with numerous black spots on the wings. Full grown larvae measured about 25 mm in length. It is brownish in colour with a pinkish tinge and numerous spiny warts on the body. The pupa is reddish brown measuring about 11 mm in length and 3 mm in breadth.

Nature of Damage

Pest appears from the flowering stage of the crop till maturity. The freshly hatched larva initially feeds on the greenish coat of capsules in between the warts. The larva then bores the capsule destroying the seeds inside. It webs together capsules along with excreta and frass. The larvae may also bore into the tender shoots and attached inflorescence killing the terminal shoots. The borer incidence usually starts from early September reaching its peak in November. The incidence declines in January.

Life Cycle

The female moth lays eggs on the tender shoots and capsules. The egg period is about 6–7 days. The larva bore into the shoots and capsules. It passes through 4–5 instars and become full fed in 2–3 weeks. Pupation takes place in a silken cocoon in shoots and capsules. The pupal stage lasts for 7–10 days. Entire life is completed in 4–5 weeks with 5–6 generations in a year. It is active throughout the year on perennial castor.

Management

1. Remove and destroy the infested shoots and capsules
2. Dusting of methyl parathion 2 per cent or chlorpyriphos 1.5 per cent or quinalphos 1.5 per cent @ 25 kg per ha should be done.

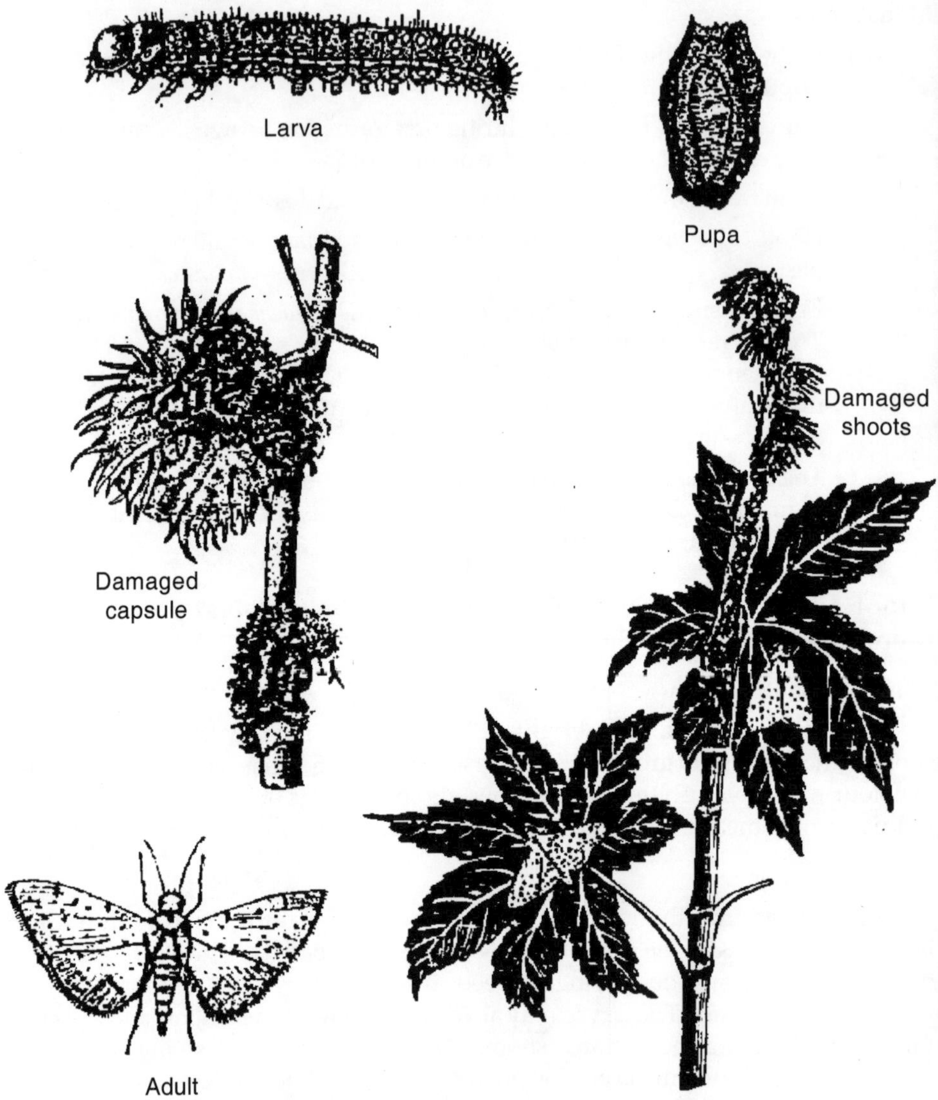

Figure 9.5: Shoot and Capsule Borer

3. Spraying of monocrotophos 0.04 per cent or chlorpyriphos 0.05 per cent should be done at an interval of 15 days for effective management of this pest.

Hairy Caterpillar, *Euproctis* Sp. (Lepidoptera: Lymantriidae)

Marks of Identification

The adult has brown forewings with dark scales and their colour extends to two spurs across yellow margin area below the apex and to the centre of the wing margin.

The Hindwings are yellow. Caterpillar is hairy dark brown with a wide yellow band dorsally on abdominal segments. A medium orange red line runs along the yellow band and a fine yellow band occurs on each side of larva above the spiracular line.

Nature of Damage

On hatching, larvae feed gregariously for sometime on host. Under severe infestation, they also feed on capsule and reduce the crop yield.

Life Cycle

The eggs are laid in masses covered with buff-coloured hairs. The egg stage lasts for 4 days. Larval development is completed in 16-20 days. Pupal stage lasts for about 8 days. The pupa is enclosed in a filmy brownish-white silken cocoon. Adult period is 2-8 days.

Management

1. Destruction of first instar gregarious larvae helps in large scale reduction in population.
2. Dusting of methyl parathion or chlorpyriphos 1.5 or quinalphos 1.5 per cent @ 25 kg per ha should be done for the effective control.
3. Spraying of monocrotophos 0.04 per cent or dichlorvos 0.05 per cent or chlorpyriphos 0.05 per cent should be done at an interval of 15 days for effective management.

Sesame or Til

Sesame Gall Fly, *Asphondylia sesami* (Diptera: Cecidomyidae)

Marks of Identification

The adults is mosquito-like and small fly. The maggots are whitish in colour.

Nature of Damage

The maggots feed on the young capsules and cause irritation to the plants, as a result, capsules are stunted, twisted and malformed. Occasionally, buds and flowers are also attacked in similar manner. The infested flowers do not develop into capsules, the pest causes maximum damage during September- October. The intensity of pest damage varied from 1-21 per cent.

Life Cycle

Female fly lays eggs in the young flower buds. The incubation period lasts from 2-4 days. Young maggots feed on the ovary of the flower. Larval period lasts for 14-24 days. The maggots pupate inside the gall. The pupal period lasts for 7-8 days. Total life cycle is completed in 23-26 days. There are 4 overlapping generations during the season.

Management

1. The adult flies can be killed by using light traps. The flies also attracted in day-time to molasses or *gur* added in water.

2. Timely sown crop can escape from the attack of the pest.
3. Infested galls should be collected and destroyed away from field to reduce further infestation.
4. Dusting the crop with methyl parathion 2 per cent or malathion 5 per cent or spray crop with endosulfan 0.07 per cent or monocrotophos 0.04 per cent.

Shoot and Leaf Webber of Sesame, *Antigastra catalaunalis* (Lepidoptera: Pyralididae)

Marks of Identification

The moth is a small insect with a wing span of about 2 cm having dark brown markings on the wing-tips. The young larvae are pale yellow and gradually become green and develop black dots all over the body.

Nature of Damage

Larvae start attack from 15thday of sowing. In 1stinstar stage, acts as leaf miner, in later stage, comes out of the mine and acts as webber by webbing the top leaves or tip of shoot, by remaining within epidermis results in drying of webbed portion. When the flowers are formed, it bores into flowers and feeds on reproductive parts. When capsule formed, it acts as capsule borer. By suitably adjusting its feeding habit, it attacks throughout the crop period. Pest is active during rainy season.

Life Cycle

Female lay up to 140 eggs singly on the tender portion of plants at night. The eggs are shiny, pale-green and they hatch in 2-7 days. The larvae feed on leaf epidermis or within the leaf tissue. It become full grown in 10-33 days and pupates in the soil in a silken cocoon. Pupal development is completed in 4-20 days. A generation is completed in about 23 days and there are nearly 14 generations in a year.

Management

1. Early sown crop *i.e.*, sown in June, is infested less than late sown crop.
2. Ichneumoids, *Temelucha biguttula*, *Eriborus* sp., *Compoplex* sp. and *Trathala flavor*-*orbitalis* are known to parasitise the larvae. *T. biguttula* is a specific larval parasite.
3. Spray the crop twice (first at pest appearance and then at flowering stage) with 250 ml of fenvalerate 20 EC.

Hawk Moth, *Acherontia styx* (Lepidoptera: Sphingidae)

Marks of Identification

The full-grown caterpillar measures about 5 cm in length and 1 cm in width, often retracts some of its anterior body segments and looks like a sphinx. The adult is a large reddish brown, robust thick-set moth with a wing span of about 10 cm. The forewings are decorated with a mixture of dark-brown and grey patterns with dark or black wavy markings and prominent yellow spots on each wing. There is a prominent

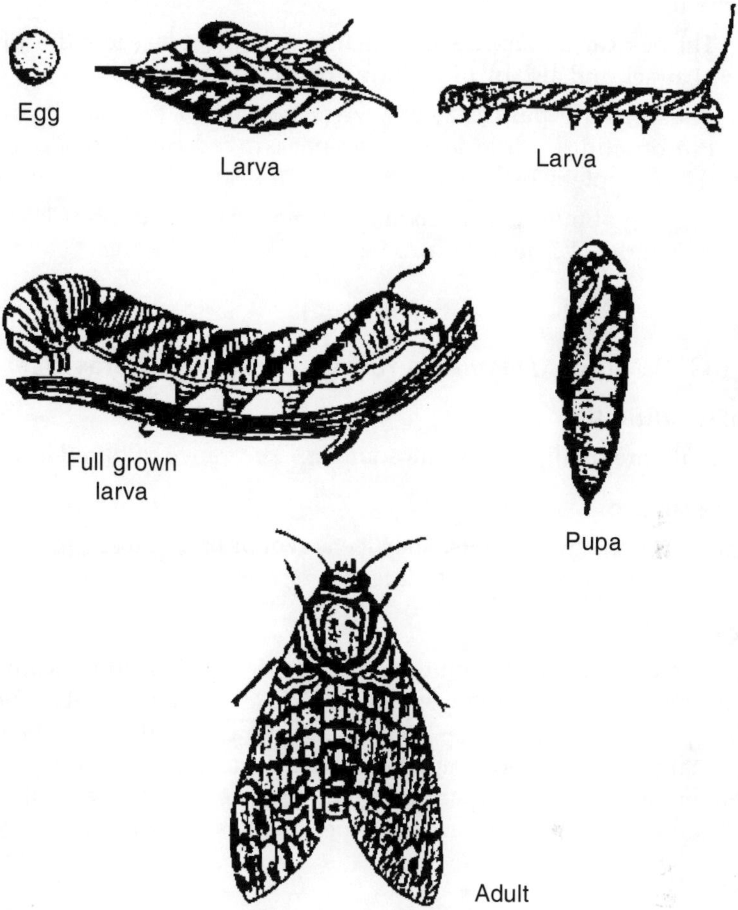

Figure 9.6: Hawk Moth

Death's head mark on the thorax. The moths are swift fliers and often make hawk like darts to a source of light at dusk.

Nature of Damage

The larvae feed voraciously on leaves and defoliate the plants. The insect is capable of inflicting heavy damage at times, but generally it is not a very serious pest in India.

Life Cycle

The female moth lay eggs singly on the underside of leaves. The eggs are fairly large in size and hatch in 2–5 days. The pale-yellow larvae start feeding on leaves. The larval period is usually long and may lasts for 2 months or more. The pupal stage lasts for 2–3 weeks in summer and about 7 months in winter. The winter is passed in pupal stage and there are 3 generations in a year.

Management

1. The pest can be suppressed by hand-picking the larvae in the initial stages of attack and also by ploughing up soil in winter season.
2. The eggs are parasitized by *Agiommatus acheronitiae* (Hymenoptera: Pteromalida) while larvae are parasitized by *Apanteles acherontiae* (Hymenoptera: Braconidae).
3. Dusting of methyl parathion 2 per cent or chlorpyriphos 1.5 per cent or quinalphos 1.5 per cent @ 25 kg per ha should be done.

Linseed

Linseed Gall Midge, *Dasineura lini* (Diptera : Cecidomyidae)

Marks of Identification

The adults are small, delicate, mosquitoes-like, orange coloured insects.

Nature of Damage

Damage is done by maggots. They feed on buds and flowers. In case of severe infestation no pod formation takes place.

Life Cycle

The females lay whitish, minute curved eggs in clusters, underneath the sepals of compact flower buds. A single female lay 22-103 eggs and infest 8-17 buds. The egg, larval, pupal and adult stages lasts for 2-5, 5-14 and 1-3 days, respectively. One generation takes 14-27 days to complete. Full-grown maggots come out of the bud fall to ground for pupation and prepare silken cocoons about 5-7 cm below the surface of the soil. There are four generations of this pest per linseed crop season.

Management

1. The adult flies should be killed by installing light trap.
2. Inter-cropping of wheat, mustard and lentil with linseed in 4:2 and 6:1 row ratio reduce the bud-fly infestation.
3. Linseed varieties *viz.*, R-552, R-958, LCK-38, EC-22582, LC-1013, LMH-412, LCK-88062, Neela and Jawahar are quite promising against the midge.
4. Spray decamethrin 0.002 per cent for effective control of this pest.

Safflower

Safflower Aphid, *Dactynotus carthami* (Hemiptera : Aphididae)

Marks of Identification

The adult is black in colour and this is one of the bigger aphid in shape and size.

Nature of Damage

In case of severe infestation the whole plant is covered with aphids, which causes excessive drainage of sap, leading to stunted growth. The yield of aphid infested crop reduced by 20-25 per cent.

Life Cycle

The duration of the life cycle is two weeks in January and one week in February. One female lays upto 13 eggs in one day. The reproduction capacity of a female is 29.5, with a maximum of 56 young ones. The life cycle completes in 8 days.

Management

1. Spray methyl demeton (0.03 per cent) or monocrotophos (0.03 per cent) or cypermethrin (0.005 per cent) or dimethoate (0.03 per cent) for control of this pest.

Safflower Capsule Fly, *Acanthiophilus helianthi* (Diptera: Tephritidae)

Marks of Identification

The adult fly is ash coloured with light brown legs. The full grown maggots are 5 mm long.

Nature of Damage

The injury is caused by the maggots which feed upon the floral parts including thalamus. The infested buds begin to rot and an offensive smelling fluid oozes at the apices giving a soaked appearance to the buds. The pest causes reduction in the yield of safflower.

Life Cycle

The adults are active from March to May. The females lay eggs in clusters of 6-24 within the flower buds or the flower. The eggs hatch in about one day and the young maggot start feeding on the florets and the thalamus. They become full-grown in one week. They pupate inside the buds. The pupal stage lasts 7 days. Three generations are completed during a crop season.

Management

1. The early removal and destruction of affected bud is helpful in checking the spread of the pest.
2. The pest can effectively and economically controlled by three sprays of dimethoate (0.03 per cent) or monocrotophos (0.03 per cent) or cypermethrin (0.002 per cent).

10

Pest of Fiber Crop

Cotton

Jassid, *Amrasca biguttula biguttula* (Hemiptera: Cicadellidae)

Marks of Identification

Adults are about 3 mm long and greenish yellow during the summer, acquiring a reddish ting in the winter. The winged adults jump or fly away at the slightest disturbance and are also attracted to light at night.

Nature of Damage

Both nymphs and adults suck cell sap from the lower surface of leaves. Affected leaves at first show signs of yellowing and curling at the margins in the lower parts of plant. Later on, crinkling and curling of leaves spread to other parts of the plant followed by bronzing and drying of leaves. Cotton jassid also causes phytotaxaemia in plants.

Life Cycle

The pest, is practically active throughout the year, but during winter, only adults were found on potato, brinjal and tomato. In spring, cotton jassid migrates to okra and then to cotton. Singh (1978) reported that adults copulated 2-10 days after maturation. They mated in a tail to tail position for 5-25 minutes. Thirty to 45 days old leaves were preferred for oviposition. Eggs were laid in the large veins on the lower surface of the leaves. A female, on an average laid 15 yellowish eggs singly during a oviposition period of 9-28 days. Egg stage lasted for 4-11 days. Nymphal stage in six instars was completed in 7-21 days. Adults lived for 5-7 weeks. There were 7-11 generations in a year.

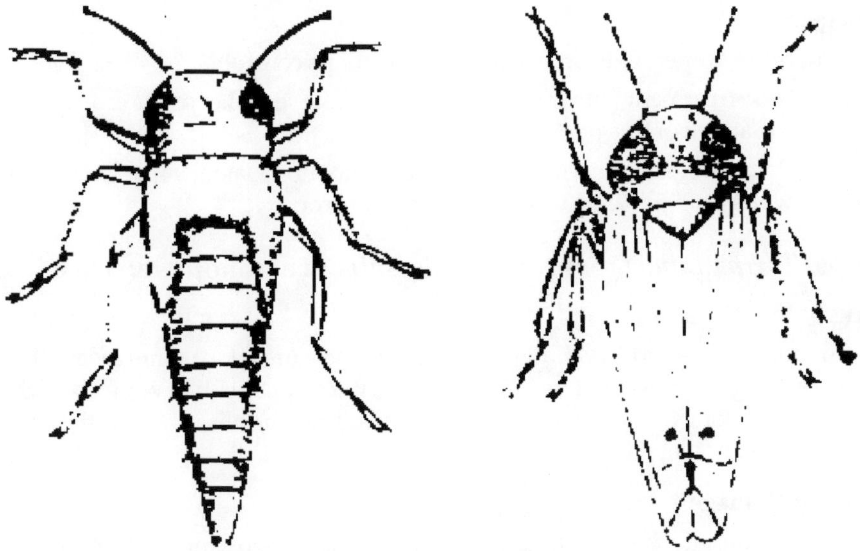

Figure 10.1: Cotton Jassid

Management

1. Use jassid tolerant varieties like L-604, LRA-5166, L-603, Savitha, Narasimha (NA-1325), NHH-44, H-8, NHH-390 and Lam Hybrid.
2. Spray with any synthetic insecticide.
3. Seed treatment with imidacloprid or carbosulfan protects the crop from all sucking pests including jassids for about a month.
4. Stem application with monocrotophos protects the crop from jassids for about a month.

Cotton Aphid, *Aphis gossypii* (Hemiptera: Aphididae)

Marks of Identification

Nymphs are light yellowish green or greenish black or brownish in colour. Adults are mostly wingless but few winged forms also seen.

Nature of Damage

Adults and nymphs both suck the plant sap gregariously. They also excrete honey dew.

Life Cycle

Pest breeds practically throughout the year on different crops. It breeds parthenogenitically and produces both alate and apterae adults. Females are viviparous. Each female produces 8-22 nymphs/day. There are four nymphal instars. Total development is completed in 7-9 days. Aphids lay eggs during winter and overwinter in this stage.

Management

1. Spray dimethoate or methyl demeton for its control.
2. Seed treatment with imidacloprid (5 g per kg seed) keeps the crop free from the attack of this pest.
3. Paint on stem with a mixture of monocrotophos: water (1:4) or imidacloprid: water (1:20) at 20, 40 and 60 days age of crop.

Thrips, *Thrips tabaci, Scirtothrips dorsalis* (Thysanoptera: Thripidae)

Marks of Identification

The adults are slender, yellowish brown and measure about 1 mm in length. The males are wingless whereas females have long, narrow strap-like wings, which are furnished with long hairs along the hind margins. The nymphs resemble the adult in shape but are wingless and slightly smaller.

Nature of Damage

Thrips attack the cotton plants at young as well as at advanced stage. At young stage, the attacked leaves become small and silvery in appearance accompanied by upward curling and crinkling. At advanced stage the affected leaves showed browning and blackening at lower surface followed by drying of leaves along the mid ribs and veins.

Life Cycle

Female lays 50-60 kidney shaped eggs singly in the tissues of tender leaves. Eggs hatch in 4-9 days. Fully fed nymphs descend to soil for pupation at depth of about 2.5cm. The nymphal, pre-pupal and pupal period last for 4-6, 1-2 and 2-4 days, respectively. Whole life cycle is completed in 11-21 days.

Management

1. Spray dimethoate or methyl demeton for its control.
2. Seed treatment with imidacloprid (5 g per kg seed) keeps the crop free from the attack of this pest.

Whitefly, *Bemisia tabaci* (Hemiptera: Aleyrodidae)

Marks of Identification

The louse-like nymphs clustered together on the under surface of the leaves and their pale yellow bodies make them stand out against the green background. In the winged stage, they are 1.0-1.5 mm long and their yellowish bodies are slightly dusted with a white waxy powder. They have two pairs of pure white wings and have prominent long Hindwings.

Nature of Damage

Adults and nymphs suck the cell sap from lower surface of leaves and cause chlorotic yellow spots on upper surface of affected leaves. Whitefly also excretes honey dew, which make the leaves sticky. Sooty mould (*Cladosporium* Sp.) growth on such leaves interferes with photosynthesis of plants.

Life Cycle

A female can lay up to 150-200 eggs. On an average, 28-43 eggs were laid singly on lower surface of leaves. Eggs are stalked, light yellow in colour and measure 0.2 mm. They hatch in 3-5 days. The nymphs on emergence feed on cell sap and grown in three stages to form the pupae within 9-14 days. The life cycle is completed in 14-122 days and 11 generations are completed in a year.

Management

1. Whiteflies can be effectively attracted and controlled by yellow sticky traps, which are coated with grease or sticky oily material.
2. Use whitefly tolerant varieties such as LPS-141 (Kanchan), LK-861 and NA-1280.
3. Spray trizophos (2.5 ml per litre of water) or profanophos (2 ml per litre of water) for effective control.

Red Cotton Bug, *Dysdercus cingulatus* (Hemiptera: Pyrrhocoridae)

Marks of Identification

The adult bugs are elongated slender insects, crimson red with white bands across the abdomen. The membranous portion of their wings, antennae and scutellum are black.

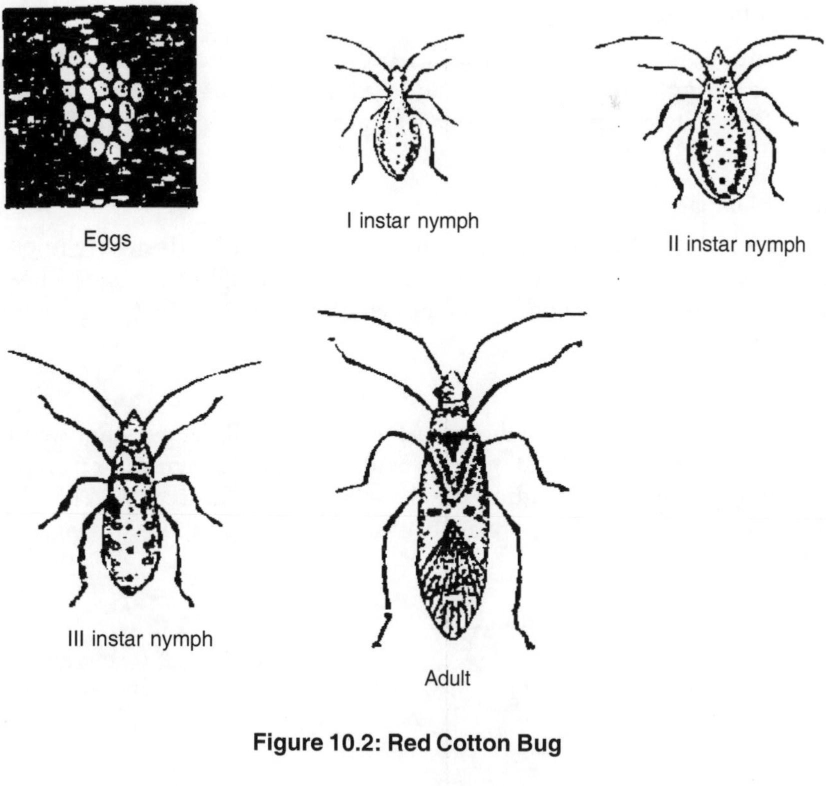

Figure 10.2: Red Cotton Bug

Nature of Damage

The damage is caused by both nymphs and adults. They suck the sap of the leaves as well as of green bolls and stain the lint by introducing a bacterium, *Nematospora gossypii*. That is way called as cotton stainer. The bugs are gregarious in habit.

Life Cycle

A female lay 100-130 eggs in loose masses in the cracks and crevices of the soil near the plant. Eggs are bright yellowish in colour. The eggs hatch in 4-6 days. The newly hatched nymphs are provided with a row of black spots and row of white spots on each side. There are 5-6 nymphal stages in the life cycle, which is completed in 19-32 days depending upon the environmental conditions. The adult live for 12-42 days.

Management

1. Spray with any synthetic insecticide.
2. Seed treatment with imidacloprid or carbosulfan protects the crop from all sucking pests including this bug.
3. Stem application with monocrotophos protects the crop from the attack of this bug.

Dusky Cotton Bug, *Oxycarenus hyalinipennis* (Hemiptera: Lygaeidae)

Marks of Identification

The adults are 4-5 mm in length, dark brown and have dirty white transparent wings. The young nymphs have a round abdomen and as they grow older, they resemble the adults, except for being smaller and having prominent wing pads instead of wings.

Nature of Damage

The nymphs and adults suck the sap from immature seeds, where upon these seeds may not ripen, may lose colour and may remain light in weight. The adults found in the cotton are crushed in the ginning factories, thus staining the lint and low its market value.

Life Cycle

The eggs are usually laid in the lint of half opened bolls, either singly or in small clusters of 3-18 each. The egg stage lasts 5-10 days and the nymphs on emergence pass through 7 stages, completing the development in 31-40 days. The life cycle completes in 35-36 days. There are many overlapping generations in a year.

Management

"Same as in case of red cotton bug."

Spotted Bollworm, *Earias vittella* and *Earias insulana* (Lepidoptera: Noctuidae)

Marks of Identification

Larvae are chocolate brown in colour and bluntly rounded. Adults of *E. vittella* are medium sized moths, head and thorax ochreous-white, forewings are pale white with a broad wedge shaped horizontal green patch in the middle, and the Hindwings are cream white in colour *i.e., E. insulana* adults are smaller than *E. vittella*. Head and thorax are pea green in colour and forewings are uniformly pale yellowish green.

Nature of Damage

When the cotton plants are young, the larvae bore into the terminal portion of the shoots, which wither away and dry up. Later on, they cause 30-40 per cent shedding of the fruiting bodies. The infested bolls open prematurely and produce poor lint, resulting in lower market value.

Life Cycle

A female lays about 400 eggs. Eggs are spherical, light bluish green in colour and sculptures, are laid singly on shoot tips, buds, flowers and fruits. Egg period is 3-7 days. Larval period lasts for 10-18 days. Full grown caterpillars are enclosed in an inverted boat shaped cocoon. Pupal period lasts for 8-12 days. Pest is active round the year and prefers high humidity and high temperature. It is more abundant in south India than North India.

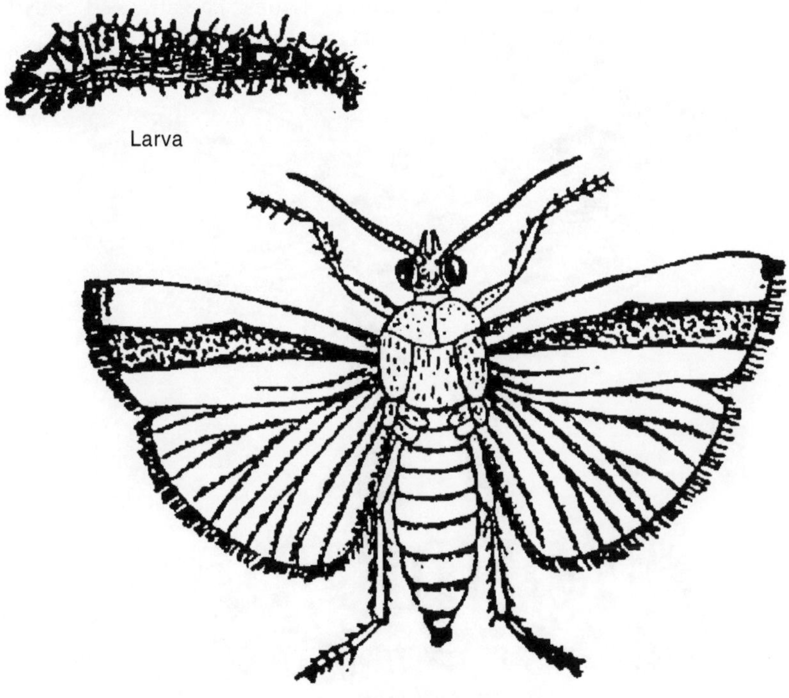

Figure 10.3: Spotted Bollworm

Management

1. All withered and drooping top shoots should be clipped early during the cotton season and destroyed.
2. Growing of a trap crop like okra helps in reducing the pest population on the cotton crop.
3. Eggs are parasitized by the hymenopteran, *Trichogramma evanescens*, and caterpillars by *Bracon* spp. and pupae by *Chelonus rufus* and *Chalcis tacharae*.
4. Application of neem products in the early stages and contact and stomach poison insecticides in the later stages particularly endosulfan (2 ml per litre) or chlorpyriphos (2 ml per ha) reduce the damage.

Pink Bollworm, *Pectinophora gossypiella* (Lepidoptera: Gelechiidae)

Marks of Identification

The caterpillars are pink in colour and found inside flower buds, panicles and the bolls of cotton. The adult moth is deep brown measuring 8-9 mm across the spread wings. There are blackish spots on the forewings, and the margins of the hindwings are deeply fringed.

Nature of Damage

The pink bollworm larvae do most spectacular damage to mature cotton bolls in which they enter as tiny just-hatched larvae, their entry holes blocked and they remain inside, devouring both seed and fibre forming tissues. The attacked bolls fall off prematurely and those, which mature do not contain good lint.

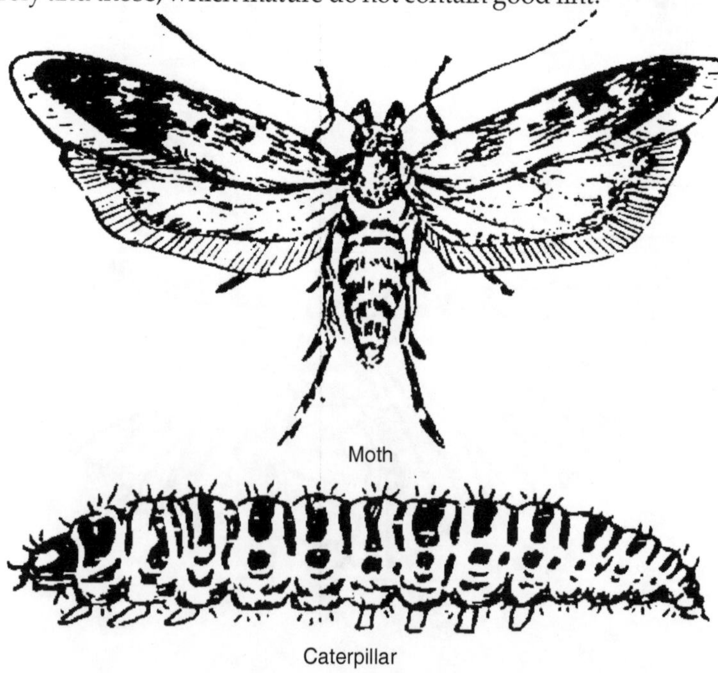

Figure 10.4: Pink Bollworm

Life Cycle

The female lay whitish, flat eggs singly on the underside of the young leaves, new shoots, flower buds and the young green bolls. The egg hatch in one week. Soon after emergence the caterpillar enters the flower buds, the flowers or the bolls. They become full-grown in about two weeks and come out of the holes for pupation on the ground, among fallen leaves, debris, etc. within one week, the moth emerge to start the life cycle all over again. By October-November, 4-6 generations are completed. The last life-cycle is very long covering 5-10 months.

Management

1. The destruction of off-season cotton sprout, alternate hosts, minimize the incidence of the pest.
2. Deep ploughing by the end of February is helpful in reducing the carry over of this pest to the next season.
3. Pest is naturally regulated by many Parasitoids *viz., Apanteles pectinophora, Bracon* spp., *Microchelonus* sp., *Elasmus* and certain predators including spider.
4. Spray endosulfan 0.07 or monocrotophos 0.1per cent or cypermethrin 0.01 per cent or monocrotophos 0.1 per cent for its control.

American Bollworm, *Helicoverpa armigera* (Lepidoptera: Noctuidae)

Marks of Identification

The moth is stoutly build and is yellowish brown. There is a dark speck and a dark area near the outer margin of each forewing. The forewings are marked with grayish wavy lines and black spots of varying size on the upper side and a black kidney shaped mark and a round spot on the underside. The hindwings are whitish and lighter in colour with a broad blackish band along the outer margin. The caterpillar when full grown is 3.5 cm in length, being greenish with dark broken grey lines along the sides of the body.

Nature of Damage

The larvae damages by boring into squares, flowers and bolls and feeds on inner contents, while feeding it thrusts it head inside the square/boll leaving the rest of its body outside. The entry hole is large and circular.

Life Cycle

The female lay eggs singly on tender parts of the plants. A single female may lay as many as 741 eggs in 4 days. They hatch in 2-6 days. The young larvae feed on the foliage for some time and later bore into the bolls and feed inside. They full fed in 13–19 days. The full grown larvae come out of the boll and pupate in the soil. The pupal period lasts 8-15 days. There may be as many as 8 generations in a year. The caterpillars feed on their fellows if suitable vegetation is not available.

Management

1. Collection and destruction of eggs and larvae on trap crop as well as main crop.

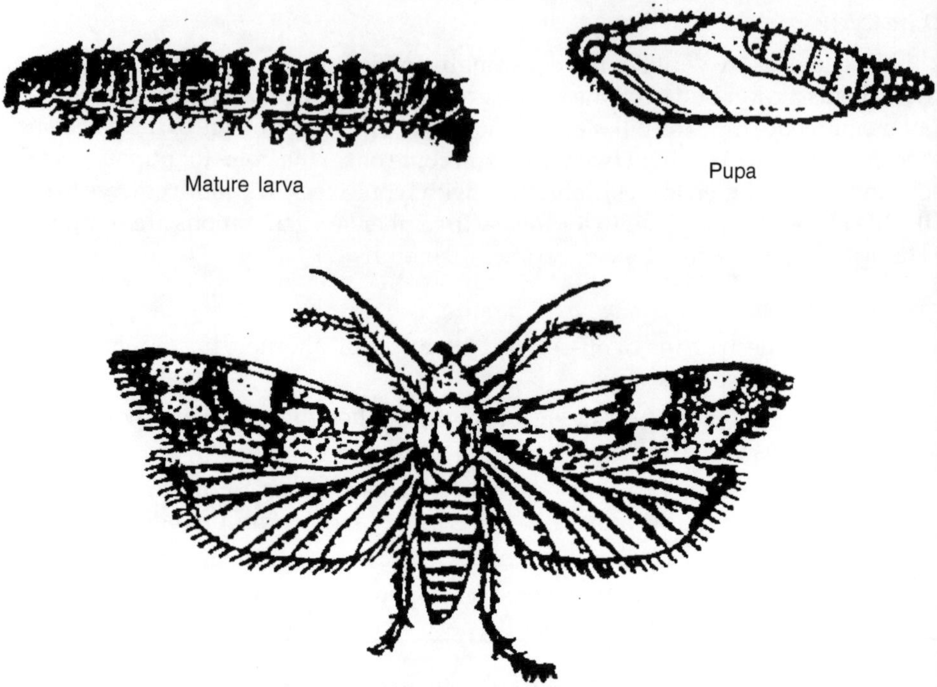

Figure 10.5: American Bollworm

2. Recommended doses of fertilizers should be applied.
3. Sow 3-4 lines of maize or jowar around the cotton crop to monitor the moth.
4. Sow trap crop like marigold 125 plant per ha along with cotton.
5. Inter crop like green gram, black gram, groundnut, soyabean should be sown.
6. Use of pheromone traps (10 trap per ha) for pest monitoring as well as trap the male moths.
7. Arrange 15-20 bird perches per ha to attract predatory birds.
8. Spraying 5 per cent neem oil before egg laying.
9. Spray monocrotophos 0.1 per cent or endosulfan 0.7 per cent or chlorpyriphos 2.0.05 per cent.
10. Among the chemicals the neem products, spinosad and emmamectin benzoate play major role in reducing the pest population.

Leaf Roller, *Sylepta derogata* (Lepidoptera: Pyraustidae)

Marks of Identification

Moths are yellowish-white, with black and brown spots on the head and the thorax. They measure about 28-40 mm across the spread wings and have a series of dark brown wavy lines on the wings.

Nature of Damage

The larvae feed on cotton leaves and in years of serious outbreaks, the cotton plants may be completely defoliated. American cotton is preferred over *desi* cotton by the pest.

Life Cycle

The pest is active from March-October and passes the winter as a full-grown caterpillar. The moths are active at night and they mate and lay 200-300 eggs singly on the underside of the leaves. The eggs hatch in 2-6 days. The young caterpillars feed on the lower surface of leaves. The older larvae roll leaves from the edges inwards upto the mid rib and feed on leaf tissues from the inside. The larvae grown through seven stages and full-fed in 15-35 days. They pupate either on the plant, inside the rolled leaves or among plant debris in the soil. They emerge as adult in 6-12 days and live for about 7 days. The life cycle is completed in 23-54 days and there are 5-6 generations in a year.

Management

1. The destruction of off-season cotton sprout, alternate hosts, minimize the incidence of the pest.
2. Deep ploughing by the end of February is helpful in reducing the carry over of this pest to the next season.
3. Spray monocrotophos (0.1 per cent) or quinalphos (0.1 per cent) or methyl parathion (0.1 per cent).

Ash Weevil, *Myllocerus* spp. (Coleoptera: Curculionidae)

Marks of Identification

The weevils are grey and are 3-6 mm long. The grubs are white, legless, cylindrical and are about 8 mm in length.

Nature of Damage

Both adults and grubs cause damage. The grub feed underground on the roots of the cotton seedlings and destroys them. One grub can destroy 9 seedlings in 40 days. The adult's feeds on leaves, buds, flowers and young bolls cut prominent round holes.

Life Cycle

A female lays 120 eggs in 24 hours. The eggs hatch in 3-5 days and the young grubs feed on the roots of the cotton and other plants. The grub completes their development in 1-2 months. They pupate into the soil inside an earthen cell. Pupal period last one week. The adult live for 8-11 days. During the active period the life cycle is completed in 6-8 weeks. There are 3-4 generations in a year.

Management

1. The pest can be suppressed by disturbing the soil upto a depth of 7.5 cm and destroying the eggs, grubs and pupae.
2. Dust 2 per cent methyl parathion (@ 20 kg per ha) or 1.5 per cent quinalphos (20 kg per ha).

11
Pest of Sugarcane

Early Shoot Borer, *Chilo infuscatellus* (Lepidoptera : Pyralidae)

Marks of Identification

The moths are light straw to brownish grey in colour. Females are slightly bigger than male moths. The wing span in male measures 19-26 mm and in female 23-35 mm. Antennae are lamellate and flat in males and filiform in females with 41 joints in both the sexes. Full grown larvae measure 20-25 mm in length and 4 mm in width. The head is dark brown and directed towards the anterior. Body is cylindrical and exhibits a dirty white colour with 5 violet dorsal strips from second thoracic to eight abdominal segments.

Nature of Damage

The borer larvae enter the plants laterally by one or more hole in the stalk and bore downwards as well as upwards killing the growing point, thereby cutting off the central leaf spindle which dries up forming a dead heart that can be pulled out easily. The cut off portion inside the bored plant rots and the dead hearts emits an offensive odour on being pulled out. The larvae feed on the soft tissues and make cavities extending to the setts. The injury to shoots after the internode formation seldom results in a dead heart. Borer infestation during the germination phase kills the mother shoots resulting in the drying up of the entire clump, creating gaps in the field.

Life Cycle

A female moth deposits about 400 eggs in one night in several egg masses. The moths live for 4-9 days. Hatching takes place in 4-6 days at sunrise or little letter. Freshly hatched larvae measure about 1.5 mm in length and have a black head and prothorax. The body is dirty grey and strips are not prominent. The larvae generally

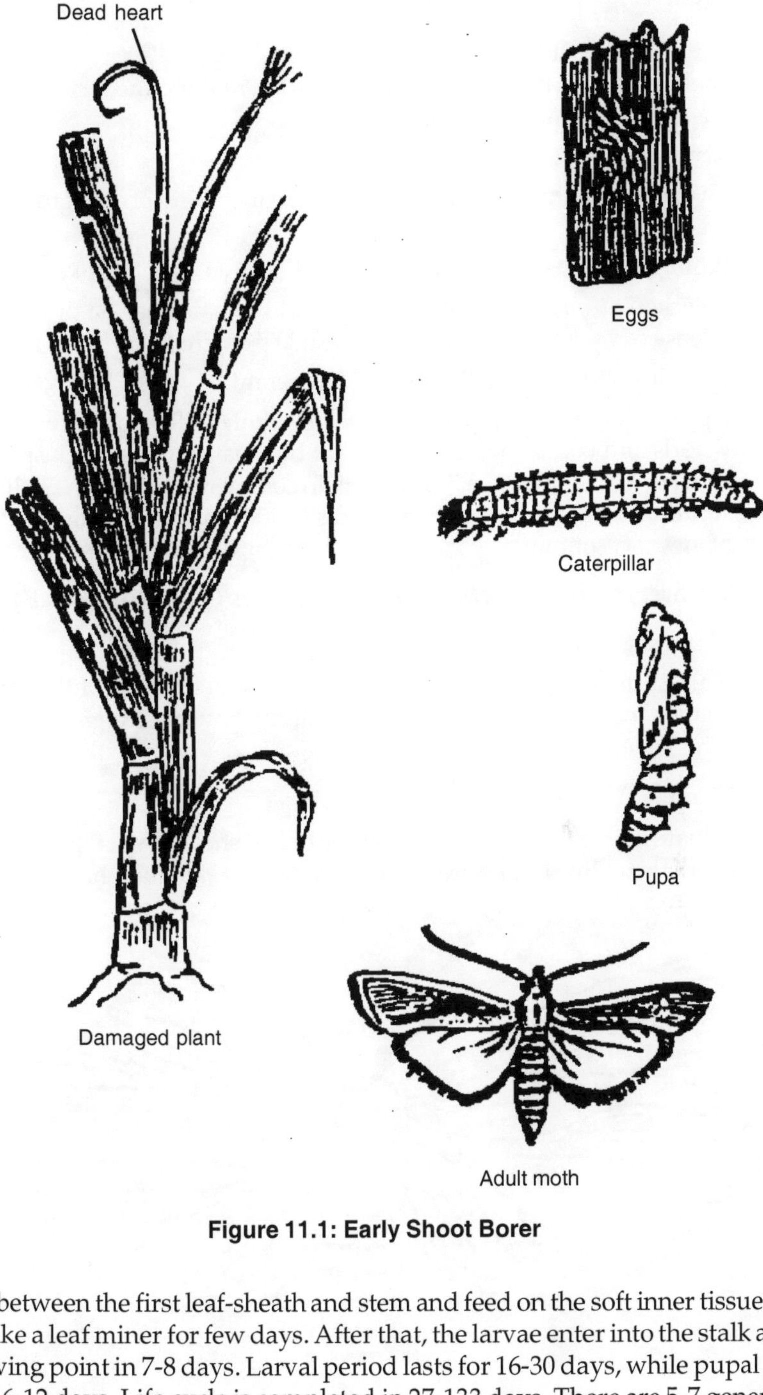

Figure 11.1: Early Shoot Borer

enter in between the first leaf-sheath and stem and feed on the soft inner tissues of the sheath like a leaf miner for few days. After that, the larvae enter into the stalk and kill the growing point in 7-8 days. Larval period lasts for 16-30 days, while pupal period lasts for 6-12 days. Life cycle is completed in 27-133 days. There are 5-7 generations in a year.

Management

1. Use of trash traps for attracting and collecting moths.
2. Socking cane setts in water for 24-48 hrs for killing the larvae within and hastening germination.
3. Removal of first leaf-sheath.
4. Two to three light earthing-up during the early stages of crop growth, reduce the incidence.
5. Pulling out of dead hearts and killing the larva with a spoke.
6. Release *Sturmiopsis inferens* @ 125 gravid females per ha or innundative release of *Trichogramma chilonis* @ 50,000 per acre.
7. Spraying shoot borer GV @ 10^7–10^9 IB per ml.
8. Application of chlorpyriphos 20 EC @ 50ml/10 litre of water on the leaf whorls and collar region of the plant by means of knapsack sprayer will prevent fur the spread of infestation. In commercial fields, the 10 litre spray solution should be applied on shoots in 100 metre row length in the point of view of economy.

Internode Borer, *Chilo (Sacchariphagus) indicus* (Lepidoptera: Pyralidae)

Marks of Identification

Moths are straw coloured with a slightly dark spot on each of the forewings. Male are smaller and darker in colour than females. They are sluggish and fly short distances when disturbed.

Nature of Damage

The neonate larvae feed on the leaf spindle or leaf sheath by scraping the tissues and characteristic white streaks are discernible on leaf lamina when it opens. The

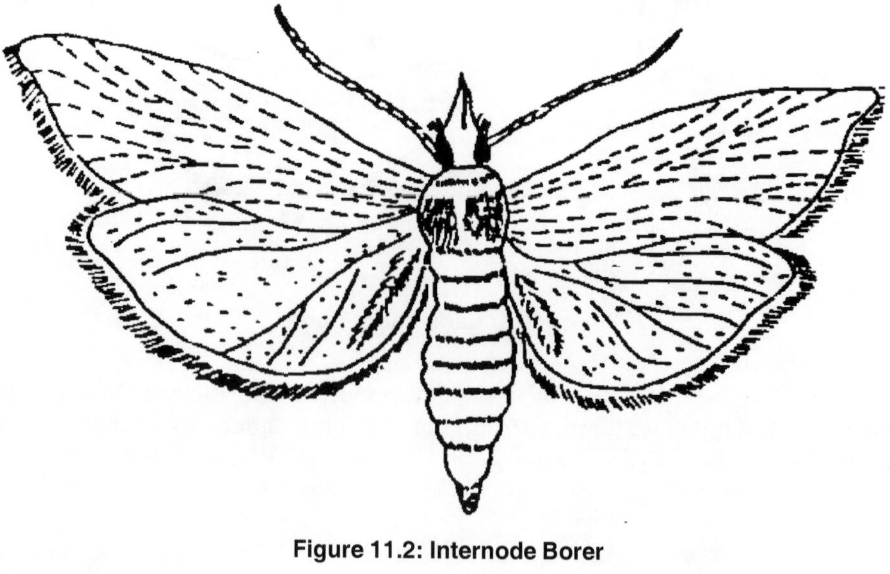

Figure 11.2: Internode Borer

larvae bore into the tender cane top. In young and weak shoots, especially in the rations, the formative internodes are damaged badly, resulting in the formation of dead hearts. There is no offensive smell when the dead hearts are pulled out. The larvae tunnel upwards in a characteristic spiral fashion, sometimes feeding extensively and at other times superficially near the periphery, depending on the variety under cultivation. Rarely it feeds downwards also. The borer feeds on the inner tissues and the frass is pushed out to the exterior. The damaged internodes get hardened and crushing of these becomes difficult.

Life Cycle

The moths lay eggs in clusters in 2–3 parallel rows on both surface of the green leaves- dorsally on the midrib and ventrally parallel to the midrib. The number of eggs in a cluster varies from a few to hundreds. The incubation period is 5–6 days. The newly hatched larvae measures about 1.5–2.0 mm in length and is light orange in colour with a black head and a prominent prothoracic shields. The larvae moult 6–7 times and complete their larval period in 37–54 days. The full grown larvae comes out of the tunnel and selects a semi-dry leaf-sheath, spins a silken cocoon in about 10–19 hours and rests inside. The pupal period ranges from 7–10 days. Under tropical conditions the borer remains active throughout the year and so all stages of the pest can be observed at any time. There 4-5 overlapping generations in a year.

Management

1. Borer free setts may be planted
2. Detrashing of the crop may be done at fifth, seventh and ninth months.
3. Water shoots may be removed at eight or ninth months.
4. High dose of nitrogen may be avoided
5. Water may be drained off in low lying areas.
6. Release *Trichogramma chilonis* @ 50000 per ha starting from the 4th to 11th month stage of the crop.
7. Install pheromone traps in water tray, for mass trapping of the males..

Top Borer, *Scirpophaga nivella* (Lepidoptera: Pyralidae)

Marks of Identification

The moth is silver white in colour. Both males and females may have one black spot on each of the forewings. The female have tuft of anal hairs of crimson red, orange or buff colour. The wing expanse is on an average 27 mm in males and 33 mm in females.

Nature of Damage

The pest occurs in all the cane growing areas in India it is easy to spot a top bored tiller with reddish brown charred and sometimes curved dead heart, leaves with shot holes and galleries in the mid-rib of the leaves showing downward passage of the borer caterpillar. In grown up canes, owing to cessation of growth, top buds

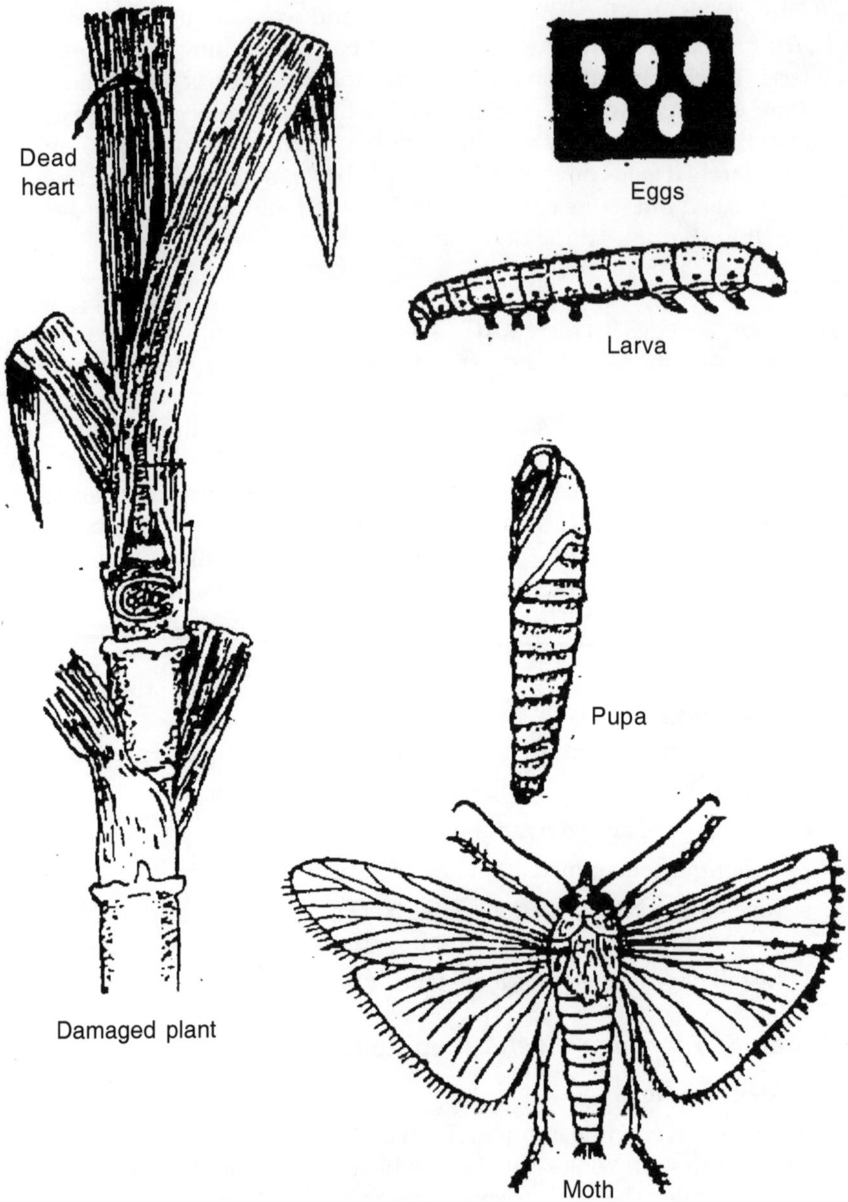

Figure 11.3: Top Borer

give rise to side shoots forming as bunchy top. The loss in yield amounts to about 20 per cent. Bored cane mature early, but with a rise in temperature from January a decrease of one to two units in sucrose per cent in juice is generally observed. Moderate and high humidity conditions favour profuse borer activity and multiplication.

Life Cycle

Eggs are laid in clusters of 2–5 overlapping rows mostly on the under surface of the leaves near the midrib and are covered by the hairs of the anal tuft. Generally, eggs are laid by a female in 3–6 masses, through rarely as high as 13 egg masses. An egg mass may have 10–80 eggs and a female lays 47–216 eggs during its life span of 1–4 days. The newly hatched caterpillars are grayish brown with black head and thorax with long hairs on the body. There are four larval instars. Upon the fourth instar, the larva deeds above the growing point. Then it cuts across the growing points, causes dead hearts, and enters the top internodes. The larvae full-fed in 21–40 days. The pupal period lasts for 4–21 days and one life cycle completed in 24–78 days.

Management

1. Collection and destruction of egg masses and moths and rouging of affected sugarcane tops before moth emergence.
2. Cut infested shoots from the top and destroy the caterpillars.
3. Carbofuran 1 kg a. i. per ha or phorate 3 kg a. i. per ha effectively control top borer and increase s the yield.

Stem Borer, *Chilo tumidicostalis* (Lepidoptera : Pyrallidae)

Marks of Identification

The full grown caterpillar measures 25-30 mm in length and 3.5 mm in width. It is white in colour with four broad pink strips present sub-dorsally and laterally in pairs. In adult moths the frons widely conical with a distinct corneous point at apex, ocelli present. Forewings are cinnamon brown, suffused with reddish to dark brown with number of scattered dark brown scales. Hindwings are whitish except for few light brown scales in the costal area in male moths.

Nature of Damage

Two types of infestation may be observes: (*i*) Primary infestation- It is caused by the newly hatched larvae aggregating in the top 3-5 internodes of the cane. As the larvae feed inside the internodes, fresh wet frass shiny red in colour is pushed to the exterior through the bore holes in the top internodes. Top leaves of infested canes dry completely. Tunneling of internodes is so severe that the dried top portion of the cane easily breaks off at the slightest jerk. (*ii*) Secondary infestation. In the phase of secondary infestation, grown-up borer larvae migrate to adjacent canes or to the lower healthy portion of the canes showing primary infestation. One caterpillar may bore one to five internodes in a cane. However, cane tops do not dry up in this case. Losses in the yield and sucrose are more due to primary than secondary infestation.

Life Cycle

the moths emerge during night and are attracted to light. Mating and oviposition take place only at night. The egg masses are deposited on the underside of the first, second and third leaves from the top in 2-4 tiers. The numbers of egg in an egg mass varies from 90-250. One female can lay as many as 800 eggs in 4-5 egg masses. The caterpillars from one egg masses hatch almost simultaneously and penetrate into one of the tender internodes. After a period of ten days, the borer larvae disperse to adjoining cans, each one boring into a separate internode. Life cycle is completed in 44-83 days with an incubation period of 7 days, larval period of 27-70 days and pupal period of 6-11 days. There is overlapping of broods and all stages of the pest are observed in the field simultaneously.

Management

1. Using light traps to collect moths.
2. Collection and destruction of egg masses
3. Removal and destruction of cane tops showing primary infestation.
4. Use of resistant varieties *viz.*, Co.-356 and Co.–513

Root Borer, *Emmalocera depresella* (Lepidoptera: Pyrallidae)

Marks of Identification

Small moth with pink head, brown thorax and abdomen. The forewings and hindwings are light yellow in colour. Forewings possess light black longitudinal stripes. Adult moth is 27 mm across the wings.

Nature of Damage

Caterpillars bore at the base of a stem, which is very close to the root. Although they do not actually bore into the roots, but since they do so near the soil surface they are called root borers. It differs from the shoot borer damage in the way that the dead-heart when pulled out the whole plant comes out and it does not possess bad smell. Besides central leaf whorl, some side leaves also dry.

Life Cycle

Moths emerge during the early morning hours and remain hidden beneath leaf-

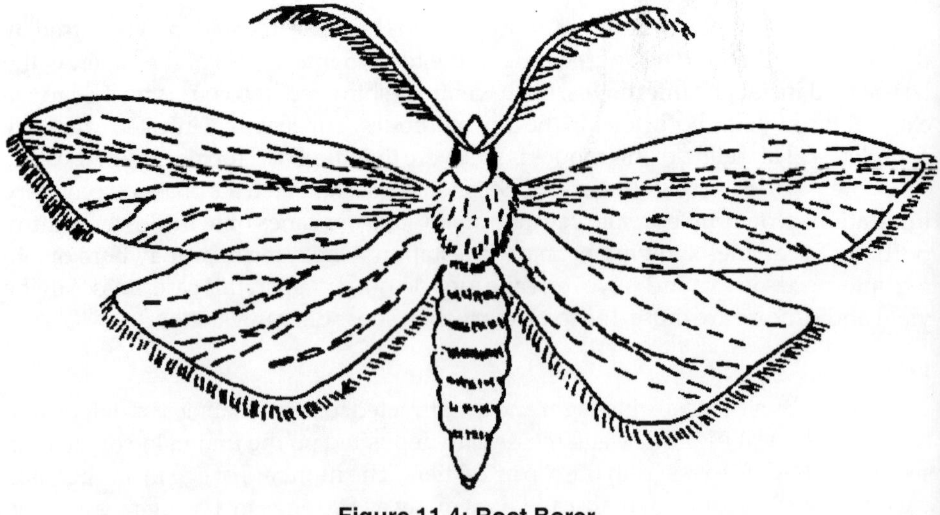

Figure 11.4: Root Borer

sheaths during the day. Mating occurs in the early hours of night, copulation lasts for 20-30 minutes. On an average, 212 eggs are laid by one female. Egg period is 5-8 days. On hatching, the larva crawls to the base of the stem and enters it by making a single hole at or just below the ground level. The larvae feed in an irregular semicircular pattern. Larval period varies between 23-43 days during hot weather, 25-59 days during rainy season and 191-245 days during winter and post monsoon period. Pupation occurs in the silken tube outside the cane. The pupal period lasts for 7-11 days during hot and rainy season and 8-14 days during winter. Duration of one life cycle is 30- 61 days in different seasons. There are 3-5 generations in a year.

Management

1. Removal of dead hearts and killing the larvae within using a pointed cycle spoke.
2. Digging and destruction of shoots at weekly intervals.
3. Collection of moths using light traps.
4. Growing twelve feet *arhar* (*Cajanus cajan*) around cane field top repel the moths.
5. Deep harvesting of cans which are to be rationed.
6. Digging and destruction of stubbles after harvest.
7. Aplication of endosulfan @1.0 kg a. i. per ha over cane setts.

The Green Borer, *Raphimetopus ablutellus* (Lepidoptera: Pyralidae)

Marks of Identification

The moths are brown in colour with a reddish tinge. The hindwings are white. Antennae of females are filamentous and those of males are comb shaped. The female moths measures 23.5 mm and male 23.3 mm. Male moths are active, while female are sluggish. The full grown larva measures 16-19 mm in length and 2-2.7 mm in width. The abdominal segments are copper-green coloured above and bluish green below.

Nature of Damage

The pest generally infests the young crop, resulting in the drying up of the central whorl to form a dead heart, which when pulled out gives an offensive smell.

Life Cycle

Eggs are deposited in a loose mass on soil and in crevices. On an average, one female lays 37 eggs. Incubation period is 3-8 days. There are five larval instars. Duration of the larval period is 20-25 days. Pupation takes place in the soil inside earthen cells. Pupal period lasts for 7-9 days. The duration of the life cycle is 30-43 days. There are three generations in a year.

Management

1. The larvae is parasitized by *Stenobracon deesae* the grub of the parasitoid feeds on the larvae.

The Gurdaspur Borer, *Bissetia stenielus* (Lepidoptera: Pyralidae)

Marks of Identification

The moth is brownish in colour with a wing span of 35-40 mm. The forewings are pale grey brown and have several blackish spots along the outer margins. The hindwings are white in colour. The larvae are polymorphic and show three colour variations *viz.*, light violet, violet and light brown and their numbers occur in the ratio of 1:2:3, respectively.

Nature of Damage

This pest attacks the plants after the cane formation. The young larvae enter the stem *i.e.*, internode and feed gregariously on the internal contents. The larva move upward in a spiral manner making minute holes on the cane surface. This is the characteristic damage due to this pest. Later on the larvae feed deeper into the stem resulting in a single straight tunnel moving upward. As a result of infestation the leaves turn yellow. The damaged internodes become weak resulting in the breaking of the stem even with slight disturbance.

Life Cycle

The eggs are laid in clusters and arranged in 304 overlapping rows in an oval pattern. The egg masses are generally deposited on the upper surface of the leaf in and along the groove of the mid rib near the leaf-sheath. The duration of egg and larval stages varies from 4–11 and 21–42 days, respectively. Pupation generally occurs within 7–8 cm of the exit hole in a specially constructed cell. The pupal period lasts for 6–13 days. The total life cycle is completed in 35–284 days. There are 2 or 3 broods in a year.

Management

1. Borer free sets may be used for planting.
2. Ratooning of heavily infested crops is to be avoided.
3. Adjusting the time of harvest also helps in checking this pest.
4. Socking setts in 0.2 per cent trichlorphon for two hours before planting kills the larvae within.

Whitefly, *Aleurolobus barodensis* (Hemiptera: Aleyrodidae)

Marks of Identification

The adult flies are pale yellow insects exhibiting brisk fluttering movement. Females are larger and less active than males.

Nature of Damage

The nymphs of white flies suck the sap from the undersurface of the leaves. As a result, the leaves turn yellow and pinkish in severe cases of infestation and gradually dry up. Heavily infested leaves are covered by the sooty mould (*Capnodium* sp.) which adversely affects photosynthesis. White fly infestation not only causes retardation in plant growth, but also reduces the sugar content in canes. Average loss in sucrose per

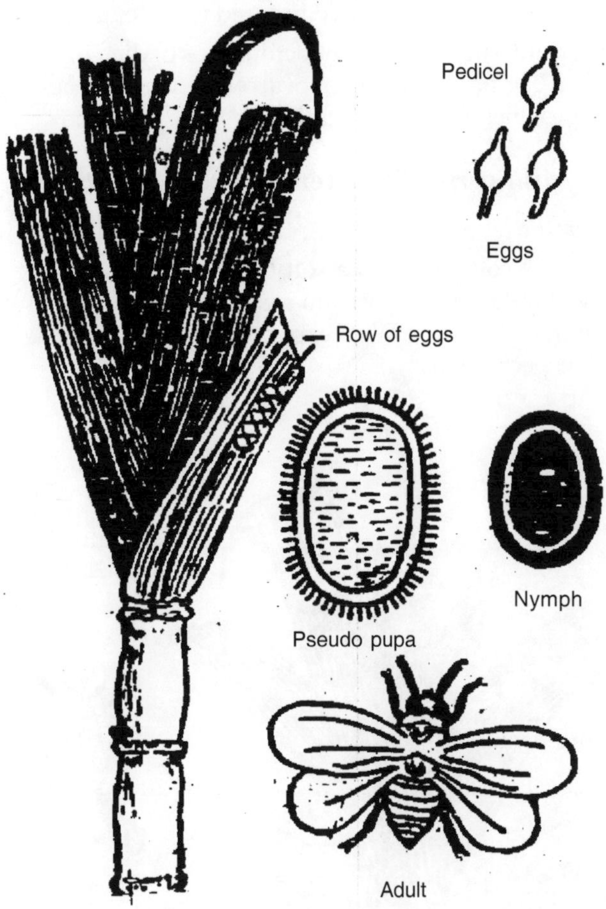

Figure 11.5: Whitefly

cent in juice amounts to 1.98 units in plant cane and 2.52-3.33 units in unmatured rations.

Life Cycle

Eggs are laid in linear rows of 2-15 mm long. Fresh egg is creamy yellow in colour, finally changing to black. Egg period lasts for 8-10 days. During winter it may prolong to 39 days. There are four nymphal instars. The duration of the first, second, third and fourth instar varies from 2-4, 4-5, 3-5 and 10-15 days, respectively. Total life cycle is completed in 25-47 days.

Management

1. Discouraging rationing in low lying areas, early harvesting of rations and adequate manuring of plant and ratoon crops with not less than 100 kg nitrogen per ha are recommended.

2. Proper drainage is to be ensured to avoid waterlogging in low lying areas.
3. Spraying with 0.04 per cent monocrotophos or dichlorovos 0.1 per cent or endosulfan 0.05 per cent after removal of dried puparia bearing lower leaves reduces pest incidence considerably.

Leaf Hopper, *Pyrilla perpusilla* (Hemiptera: Fulgoridae)

Marks of Identification

The adult is straw coloured, has a soft body and is very active. The head is prominently drawn forward into a rostrum or snout and measures 10 mm in length. Females are longer than the males. Wing expanse of male and female varies from 16-18 and 19-21 mm, respectively.

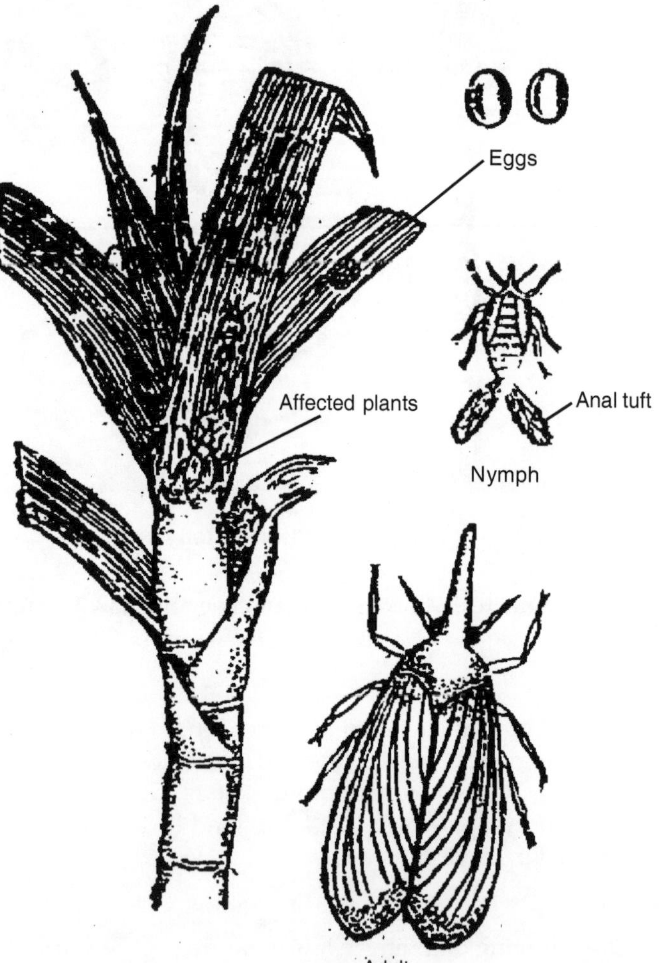

Figure 11.5: Leaf Hopper

Nature of Damage

The adults as well as the nymphs suck the sap from the undersurface of the lower leaves near midrib, resulting in yellowish white spots. When the infestation is high, the leaves gradually turn pale and wither away. The hopper exude a sweet sticky fluid commonly known as *honey dew,* which promote quick and luxurious growth of the fungus, *Capnodium* sp., and as a result the leaves get completely covered by the sooty mould. The black coating interferes with the photosynthesis of the leaves and crop growth is adversely affected. Due to continuous desapping by large number of hoppers, top leaves in the affected canes dry up and the lateral buds germinate.

Life Cycle

Eggs are laid in clusters of 30-50, covered with waxy threads on the ventral side of green leaves parallel to the midrib. Each female lays 600-800 eggs. The female generally prefers low, shady and concealed sites for oviposition. Incubation period during summer, monsoon and winter months ranges from 10-15, 6-10 and 15-18 days, respectively. There are five nymphal instars. The duration of each instar is 7-10 days. The total nymphal period during summer, monsoon and winter months ranges from 40-52, 34-40 nd 70-132 days, respectively. Females live longer (6-8 weeks) than males (4-6 weeks). The pest completes 4-5 generations in a year.

Management

1. Burning of trash after harvest should be carried out upon the middle of March in order to destroy the unhatched egg masses and over-wintering nymphs.
2. Removal of sprouts from the stubble at least once by the end of April helps in rducing the pest build-up.
3. Removal of dry leaves from August onwards, effectively reduces pyrilla population. During pre-monsoon period, eggs are laid on the lowest two leaves which may be removed and destroyed.
4. Spraying with *Metarhizium anisopliae* @10^7spores per ml.
5. Release of pyrilla adults seeded with *Metarhizium anisopliae* spores @ 250 adults per ha.
6. Redistribution and colonization of *Epiricania melanoleuca* cocoons/egg masses from heavily parasitized fields to low parasitized field.
7. Spraying with monocrotophos (0.04 per cent) or endosulfan (0.05 per cent) after removal of dried lower leaves reduces pest incidence considerably.

12

Pest of Vegetables

Brinjal

Shoot and Fruit Borer, *Leucinodes orbonalis* (Lepidoptera: Pyralidae)

Marks of Identification

Adults has brownish and red markings on the whitish forewing. Hindwings are opalescent with black dots.

Nature of Damage

It is the most important and destructive pest of brinjal. It starts damaging brinjal plant, a few weeks after transplanting, larva bores into tender shoots and causing withering of terminal shoots. It also bores into petiole of the leaves, flower buds and developing fruits causing withering of leaves, shedding of buds and making the fruits unfit for consumption and marketing. Attacked fruits show holes on them plugged with excreta. It cause 70 per cent loss.

Life Cycle

Eggs are laid singly on leaves, shoots and fruits. They hatch in 3-5 days. Larva is a borer within shoot, leaf midrib, petiole and fruit and feeds on the internal tissues. It become full fed in 10-15 days. It is stout and pink coloured with sparsely distributed hairs on warts on the body and brownish head. Pupation takes place in a tough greyish cocoon on the plant itself for a period of 6-8 days. Adult lives for 2-5 days and female lays upto 250 eggs.

Management

1. *Destroying Brinjal Crop Stubble*

Many farmers store dried Brinjal stubble from the previous season crop for use as fuel for cooking. Such stubble is heaped around the field or near by dwelling. Fruit

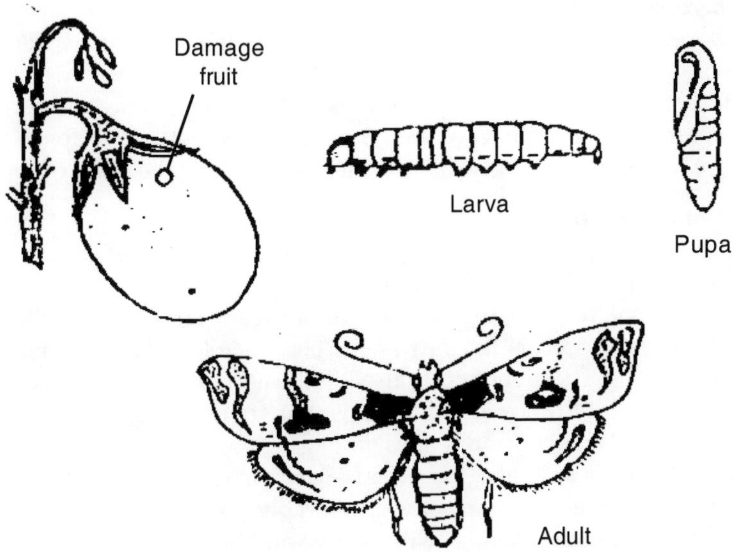

Figure 12.1: Brinjal Shoot and Fruit Borer

and shoot borer pupae can survive in this stubble for several weeks. Then when the new brinjal crop is planted, the moths from these pupae fly and lay eggs on the crop, starting a pest epidemic. This stubble should be destroyed, either burned or buried after harvest.

2. Use Healthy Pest Free Seedlings

Most farmers grow brinjal seedling in the open field, often near abandoned brinjal crops or heaps of dried brinjal stubble from the previous season. Adults from these old plants fly and lay eggs on seedlings. These eggs are very hard to see. The use of such contaminated seedlings spreads the pest into the production field. To avoid this, brinjal seedlings should be raised away from source of infestation. Also growing seedlings under nylon netting prevents borer moths from laying eggs on the plants.

3. Remove and Destroy Infested Shoots

Before plant start fruiting the larvae feed inside the tender shoots. These are visible as dried tips. Cut and destroy these larvae infested shoots immediately. This will helpful in reducing the pest population. These shoots must be destroyed by burning or bury them at least 20 cm deep in soil. These pruning activities are very important in the early season. Once fruiting begins, most larvae will prefer to enter in fruits rather than shoots. In newly infested fruits, it is very difficult to detect the damage and the insect population multiply and spread the infestation. Always destroy damaged or infested fruits. After the final harvest, the old plants should be uprooted and destroyed promptly because they may harbor pest larvae, which could become a source of infestation in future.

4. *Use of Pheromone Traps*

The pheromone lures are commonly available. The pheromone traps will be installed in the field, these can attract male moths continuously for up to six weeks. Traps should be installed in the field 2-3 weeks after transplanting and continue till the last harvest. 10-15 m distance should be maintained between traps in the field. The traps are hung in such a way that the lure is just above the plant canopy.

5. *Use of Chemical Pesticides*

The indiscriminate use of toxic, broad spectrum insecticides should be avoid. This is because with in hours of hatching from eggs the larvae enter fruit or shoot. Once inside these plant parts, insecticides cannot reach larvae and kill them. Also due to frequent use of insecticides, the pest has become tolerant and cannot kill easily with these chemicals. These chemicals also kill the natural enemies such as spiders, mantis, earwigs, ladybird beetles and wasps that are naturally found in the field. These predators feed on pests and reduce damage to the brinjal crop. For this reason, the use of pesticides, especially the broad spectrum once has to come down for successful control of this pest. If selective, preferably, biological insecticides, such as neem are used the natural enemies will survive and be able to help kill fruit borer larvae. If it should be necessary then always apply least toxic insecticides at the recommended dose.

Stem Borer, *Euzophera perticella* (Lepidoptera: Pyralidae)

Marks of Identification

The moth measures about 32 mm across the spread wings and have pale yellow abdomens. The head and thorax are grayish, the forewings are pale straw-yellow and the hindwings are whitish. The full grown caterpillar is creamy white and have a few bristly hairs. Their body taper posteriorily.

Nature of Damage

Larvae bore into main stem of both young and old plants and moves downward. Top shoots of young plants droop and wither and older plants become stunted and fruit bearing capacity is adversely affected.

Life Cycle

Eggs are laid singly or in batches on young leaves, petioles and tender branches. They hatch in 3-10 days. Yellowish larvae with red head feeds on the exposed parts for a few minutes after which it bores into the stem at leaf or branch axil and cover the holes with excreta and frass. It feeds on the internal tissues and becomes full grown in 26-58 days. Pupation is within a silken cocoon inside. Pupal period lasts for 9-16 days. Total life cycle ranges from 26-58 days.

Management

1. Collect and destroy affected plants.
2. Avoid continuous cultivation and rationing of brinjal.
3. Spray endosulfan 35 EC @ 2 ml with neem oil @ 2 ml per litre starting from one month after planting at 15 days interval.

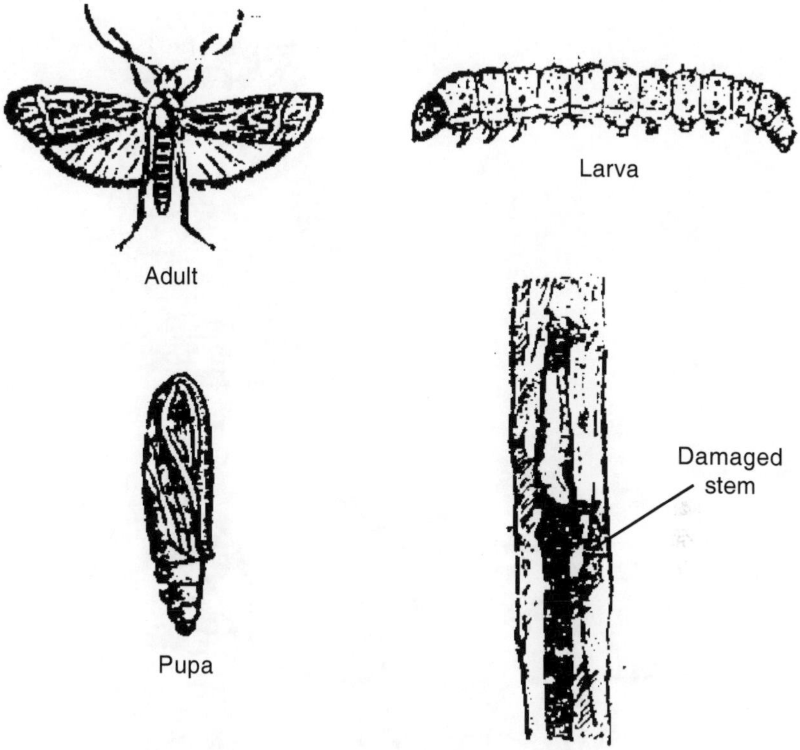

Figure 12.2: Brinjal Stem Borer

Hadda Beetle, *Epilachna viginitioctopunctata* and *E. dodecastigma* (Coleoptera: Coccinellidae)

Marks of Identification

Adults are hemispherical in shape, pale brown in colour and mottled with black spots. *Epilachna viginitioctopunctata* has 14 spots on each elytra, while *E. dodecastigma* has six spots.

Nature of Damage

Both adults and grubs cause considerable damage to the leaves by scraping away chlorophyll from epidermal layers of leaves which get skeletonized and gradually dry away.

Life Cycle

A female lays 120-180 eggs. Egg period is 2-4 days. Grubs are yellowish in colour, stout with spines all over the body. Larval period lasts for 10-35 days and pupation occurs on the leaves or stem. Pupa is yellowish with spines on the posterior part and the anterior portion being devoid of spines. Pupal period lasts for 3-5 days. Life cycle takes 20 to 50 days to complete and depends upon weather conditions. There are 7 generations in a year.

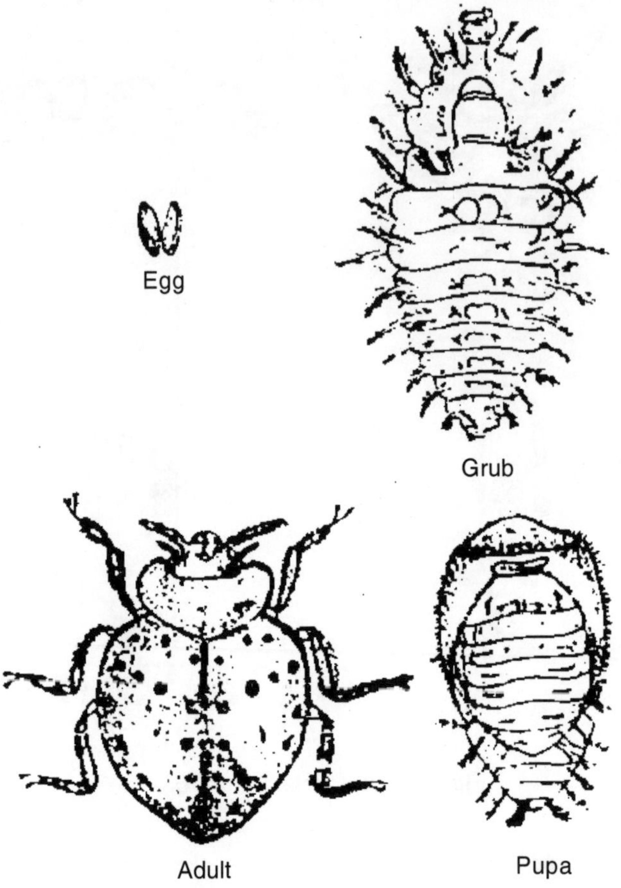

Figure 12.3: Hadda Beetle

Management

1. Collect damaged leaves and egg masses on the leaves and destroy them.
2. Shake the plants to dislodge grubs, pupae and adults in a pail of kerosenated water early in the morning or collect them mechanically and destroy.
3. Spray endosulfan 35 EC 2ml per litre or malathion 50 EC 2 ml per litre of water.

Tomato

Fruit Borer, *Helicoverpa armigera* (Lepidoptera: Noctuidae)

Marks of Identification

Adult are a medium sized light brown stout moth with a "V" shaped speak on the forewing. Hindwings are with dull black border. The young caterpillars are yellowish-white in colour and gradually become greenish in colour.

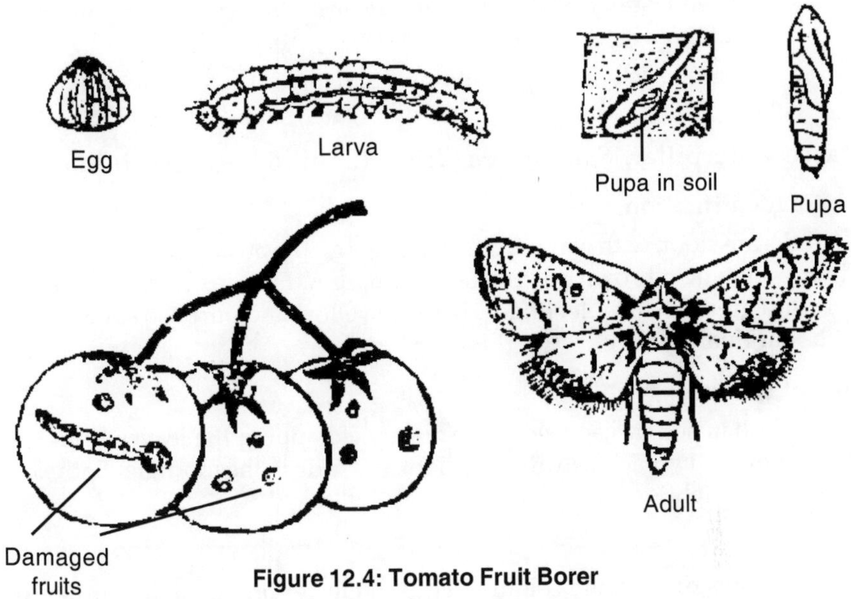

Figure 12.4: Tomato Fruit Borer

Nature of Damage

The young larvae feed on tender foliage and full grown caterpillar attack fruits. They make circular bore holes and thrust only a part of their body inside fruit and eat inner contents. Caterpillar move from one fruit to another and one caterpillar consume 2–08 fruits in its life span.

Life Cycle

Eggs are yellowish white, ribbed and dome shaped and laid singly on leaves and flowers. Egg period lasts for 2-4 days. Larval period lasts for 15-24 days. Full grown larvae drop down from the plants and burrow into the soil and pupate therein. Pupal period is 10-14 days. Total life cycle completed in 4-6 weeks. There are 5-8 generations in a year.

Management

1. Grow less susceptible genotypes *viz.,* Rupali, Roma and Pusa red palm.
2. Grow simultaneously 40 days old American tall marigold and 25 days old tomato seedlings at 1:16 rows to attract the female moth for egg laying.
3. Set up pheromone traps with *heli* lure at 15 per ha and change the lure once in 15 days.
4. Collect and destroy the damaged fruits and grown-up caterpillars.
5. Six release of *Trichogramma chilonis* @ 50,000 per ha per week coinciding with flowering time and based on ETL.
6. Spray *Helicoverpa armigera* NPV at 500 LE per ha along with cotton seed oil (300 g per ha), to kill larvae.

7. Spray endosulfan 35 EC 2 ml per litre or *B.t.* 2 g per litre or quinalphos 2.5 ml per litre.
8. Encourage activity of parasitoids, *Eucelatoria bryani, Campolites, Chelonus,* etc.

Tobacco Caterpillar, *Spodoptera litura* (Lepidoptera: Noctuidae)

Marks of Identification

Adults are stout, with wavy white markings on the brown forewings and white hindwings having a brown patch along its margin. Full grown caterpillar is pale brown with greenish to violet tinge. There are yellow and purplish spots present in the sub-marginal areas.

Nature of Damage

Freshly hatched larvae feed gregariously, scrapping the leaves from ventral surface. They feed voraciously during night and hide in the morning. Entire crop is defoliated overnight.

Life Cycle

Female lays dirty white coloured eggs in cluster on the under surface of the leaves and covered with brown hair. Incubation period is 3-5 days. Larval period is 20-28 days. Pupation takes place in the soil earthen cocoon for 7-11 days. Total life cycle is completed in 30-40 days during summer and 120-140 days in winter season. There are 7-9 overlapping generations in a year.

Management

1. Plough the soil to expose and kill pupae.
2. Grow caster along border and irrigation channels as indicator or trap crop.
3. Flood the field to draw the hibernating larvae.
4. Set up light or pheromone traps as 15 per ha.
5. Remove and destroy egg mass in castor and tomato.
6. Collect damaged leaves and early instar gregarious larvae and destroy.
7. Hand picking of grown-up larvae and kill them.
8. Spray NPV for *S. litura* at 250 LE along with teepol 1 ml per ha in evening hours.
9. Spray chlorpyriphos 20 EC 2.0 litre per ha or DDVP 76 WSP 1.0 litre per ha or endosulfan 35 EC 1.25 litre per ha or NSKE 5 per cent.
10. Prepare poison bait with rice bran 5 kg, jaggery 0.5 per cent, carbaryl 50 WP 0.5 per cent, water 3 litre per ha and spread the bait in the evening hours.

Hadda Beetle, *Epilachna viginitioctopunctata* and *E. dodecastigma* (Coleoptera: Coccinellidae)

"As described in Brinjal pests."

Whitefly, *Bemisia tabaci* (Hemiptera: Aleyrodidae)

Marks of Identification

The adults are tiny white in colour, usually crowd in between the veins on ventral sides of leaves. Nymphs are oval, scale like and greenish white in colour.

Figure 12.5: Whitefly Affected Plant

Nature of Damage

Nymphs and adults suck the sap from the tender leaves. The affected leaves become yellow, the leaves wrinkle and curl downward and are ultimately shed. These insects also exude honey dew which favour the development of sooty mould. These insects also act as vector, transmit leaf curl virus disease and cause severe loss.

Life Cycle

Eggs are pear shaped, light yellowish in colour and are anshores upright on leaves by tail like appendages. Incubation period is 3-5 days. On hatching the nymphs suck the sap. Nymphal development takes place in 9-14 days in summer and 20-80 days in winter. Pupal period lasts for 3-8 days. Total life cycle is completed in 14-107 days. There are about 9-11 overlapping generations in a year.

Management

1. Uproot and completely destroy the diseased leaf curl plants.
2. Collect and destroy the damaged leaves along with eggs, nymphs, pupae and adults.
3. Remove alternate weed hosts.
4. Use nitrogen fertilizers and irrigation judiciously.
5. Use yellow sticky traps smeared with sticky substrates to attract and kill the adults.
6. Spray fish oil rosin soap (FORS) 2 per cent or neem oil 0.5 per cent along with teepol 1 ml per litre or methyl demeton 0.025 per cent or endosulfan 0.07 per cent along with FORS.
7. Apply systemic insecticides in early stage of the plant growth and contact insecticides in the later stages for vector control.
8. Encourage activity of parasitoids, *Eretmocerus masii* and predatory coccinellid *viz., Brumus* and *Chrysoperla*.

Okra

Shoot and Fruit Borer, *Earias vittella, E. insulana* (Lepidoptera: Noctuidae)

Marks of Identification

Larvae are chocolate brown in colour and bluntly rounded. Adults of *E. vittella* are medium sized moths, head and thorax ochreous-white, forewings are pale white with a broad wedge shaped horizontal green patch in the middle, and the hindwings are cream white in colour *i.e., E. insulana* adults are smaller than *E. vittella*. Head and thorax are pea green in colour and forewings are uniformly pale yellowish green.

Nature of Damage

Caterpillar bore into tender shoots and tunnel downwards. Shoots wither, droop and growing points are destroyed. Caterpillar also bore into buds, flowers and fruits and feed on inner tissues. Damaged buds and flowers wither and fall down. Damaged fruits become deformed in shape and remain stunted in growth.

Life Cycle

A female lays about 400 eggs. Eggs are spherical, light bluish green in colour and sculptures, are laid singly on shoot tips, buds, flowers and fruits. Egg period is 3-7 days. Larval period lasts for 10-18 days. Full grown caterpillars are enclosed in an inverted boat shaped cocoon. Pupation takes place in 8-12 days. Pest is active round the year and prefers high humidity and high temperature.

Management

1. Adopt clean cultivation and remove alternate host plants.
2. Do not cultivate cotton and okra together or one after other in a year to check the population build-up of pests.
3. Collect and destruction of affected fruits.
4. Grow resistant/tolerant varieties *viz.*, Kalyanpur Boni, Pusa Makhmali, HB-55, AC-218.
5. Use pheromone traps for *Earis* spp.
6. Release of egg parasitoids *Trichogramma chilonis* at 100000 per ha and egg larval parasitoids *Chelonus blackburni* at 10000 per ha.
7. Spray endosulfan 35 EC 750 ml per ha or monocrotophos 36 WSC 650 ml per ha or B.t.2 g per litre of water.

Leaf Hopper, *Amrasca bigutella bigutella* (Hemiptera: Cicadellidae)

Marks of Identification

Adults are wedge shaped, pale green in colour with black dot on posterior portion of each forewing.

Nature of Damage

Both nymphs and adult suck cell sap from the ventral surface of leaves and inject their toxic saliva into plant tissues. Infested leaves turn yellowish and curl. In case of severe infestation, leaves turn dark brick red, become brittle and crumple.

Life Cycle

Pear shaped, elongated and yellowish white eggs are laid singly in the tissues of main veins on the undersurface of leaves. Egg period lasts for 4-10 days. Newly emerged nymphs suck the sap from tender portion and leaves of plant and become adult in 7-20 days. Pest population appears with the onset of cloudy weather and decreased after heavy monsoon showers.

Management

1. Uproot and completely destroy the diseased plants.
2. Collect and destroy damaged leaves along with eggs, nymphs and adults.
3. Remove alternate weed hosts.
4. Use yellow sticky traps smeared with sticky substances to attract and kill adults.
5. Grow resistant/tolerant varieties *viz.*, AE-30, Reshmi, AC-3375, Okra Red, Sel.- 6,8,11,HR-4,Punjab-7.
6. Spray fish oil rosin soap (FORS) 2 per cent or neem oil 0.5 per cent along with teepol 1 ml per litre or methyl demeton 0.025 per cent or endosulfan 0.07 per cent along with FORS.
7. Apply systemic insecticides in early stage of the plant growth and contact insecticides in the later stages for vector control.

Cucurbitaceous Vegetables

Fruit Fly, *Bactrocera cucurbitae, B. ciliatus* (Diptera: Tephritidae)

Marks of Identification

This fruit fly distributed throughout the country and attack almost all kinds of vegetable and a very serious pest. Adult is reddish brown fly with lemon yellow surved vertical markings on thorax and fuscous shading on outer margins of wings. *B. ciliatus* it is a polyphagous attacking all kind of all kinds of cucurbits. Adult fly is bright brown or ferrugineus brown with hyaline wings and two round dark brown spots on fourth abdominal segment.

Nature of Damage

Maggots tunnel and feed within fruits and cause damage. Infested fruits decay aided by bacterial action. They rot and drop down. Fly prefers tender fruits. Adult also causes injury by making oviposition puncture on the fruits through which fruit juice oozes out. Above 50 per cent loss is caused to vegetables by this fly.

Life Cycle

Female flies puncture soft and tender fruits and lay eggs. Egg are laid in the fruit

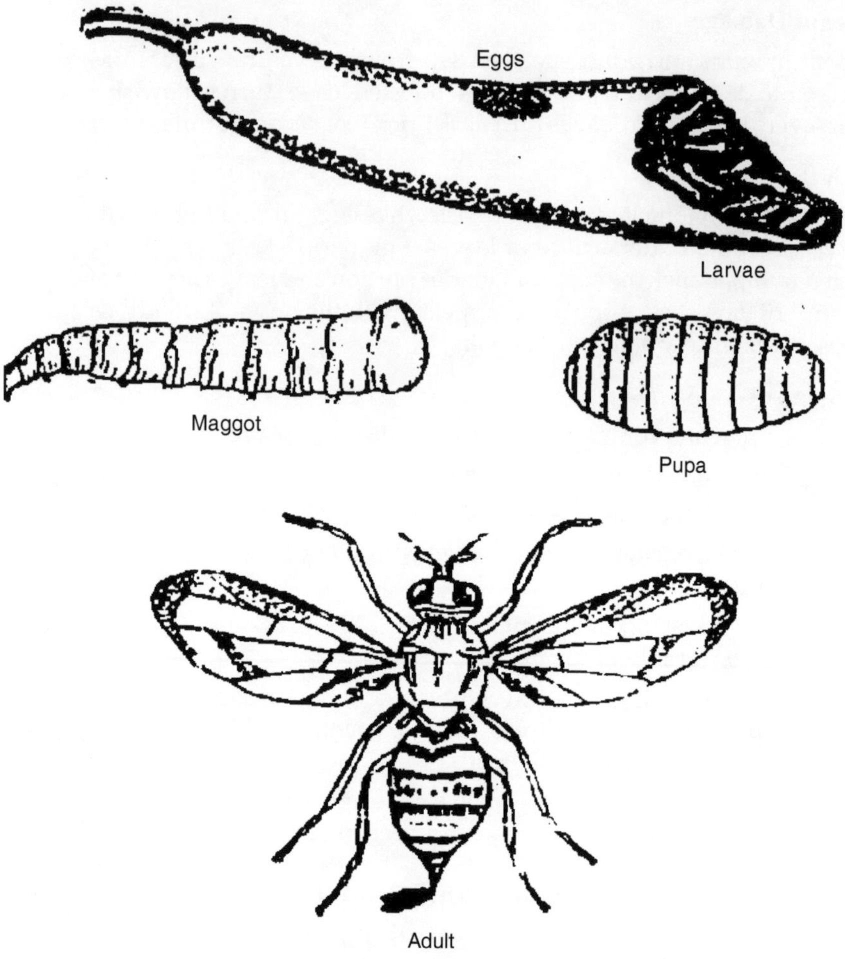

Figure 12.6: Fruit Fly

in a cavity 2-4 mm deep, singly or in clusters of 4-10 and sealed with a gummy secretion from ovipositor. Oviposition takes place at intervals of 2-5 days, laying 1-26 eggs a day. Egg period lasts from 1 day in summer to 6-9 days in winter. Larva bores into fruit feeding on the internal contents. Larval period varies from 3 days to 3 week. Fully grown larvae falls to ground to pupate in soil. Pupal period is 6-15 days. Male fly live up to 56 days and the female up to 66 days.

Management

1. Grow resistant or early maturing variety (Arka-tinda, a variety of round gours, Arka-saryamakhia- pumpkin variety)
2. Change sowing dates as the fly population is low in hot dry conditions and its peak during rainy season.

3. Use attractants like citronella oil, eucalyptus oil, vinegar, dextrose and lactic acid to trap flies.
4. Use ribbed guard as trap crop and apply carbaryl 0.15 per cent or malathion 0.1 per cent on congregating adult flies on the under surface of leaves.
5. Collect and destroy all fallen and damaged fruits by dumping in a pit and covering with a thick layer of soil to prevent carry over of pest.
6. Frequently rake-up soil under the vines or plough infested field to destroy puparia and apply endosulfan 4 per cent dust at 25 kg per ha on the soil.
7. Use methyl eugenol lure (25 per ha) to monitor and kill adults of fruit flies or prepare methyl eugenol and malathion 50 EC mixture at 1:1 ratio and take 10 ml mixture with six holes and place 12 traps per ha or keep 5 g wet fish meal in a polythene bag with six holes and add 0.1 ml of dichlorvos and place 12 traps per ha.
8. Protect fruits with polythene bags.
9. Use bait spray combining molasses or jaggery 10 g per litre and one of the insecticides, fenthion 100 EC 1 ml per litre or malathion 50 EC 2 ml per litre or dimethoate 30 EC 1 ml per litre, two rounds at fortnight interval before ripening of the fruits.

Pumpkin Beetle, *Aulacophora foveicollis* (Coleoptera: Chrysomelidae)

Marks of Identification

Adult is oblong. Dorsal surface is orange-red while ventral surface is black. Grub is creamy white with a silghtly darker oval shield at the back.

Nature of Damage

Damage is caused mainly by adult beetles, which feed extensively on leaves, flowers and fruits making holes and cause death or retardation of growth. Seedlings, when infested are totally destroyed. Damage done by grub to the seedling also is serious.

Life Cycle

Red coloured beetle lays spherical eggs singly or in batches in the moist soil around the base of the host plants. As many as 300 eggs are laid by a female. Egg period lasts for 6-15 days, larval period for 13-25 days and pupal period for 7-17 days. Grub undergoes four instars entering soil each time to moult. Mature larvae enter soil and pupate within a water proof cocoon. Five generations are completed in a year. Adults live for more than one month. They hibernate under old cucurbitaceous creepers, grasses and weeds and in soil.

Management

1. Ensure clean cultivation.
2. Adopt early sowing and cultivate less susceptible varieties.

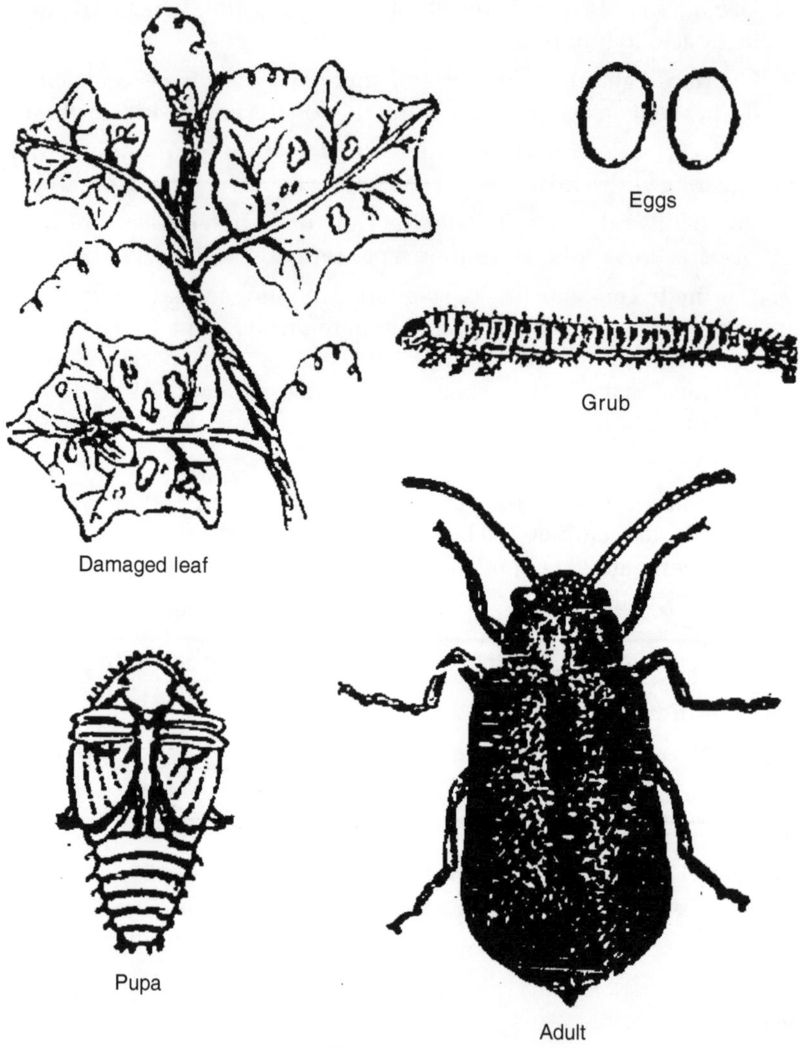

Figure 12.7: Pumpkin Beetle

3. Collection and destruction of adult beetles.
4. Spray malathion 50 EC or dimethoate 30 EC 1 ml per litre or fenthion 100 EC 1 ml per litre.

Cruciferous Vegetables

Diamond Back Moth, *Plutella xylostella* (Lepidoptera: Plutellidae)

Marks of Identification

Adult is a small grayish moth, which when at rest shows a series of three yellowish diamond, shaped markings dorsally on the wings. Larvae feeds on foliage and grows to a pale-green caterpillar, brownish at the anterior.

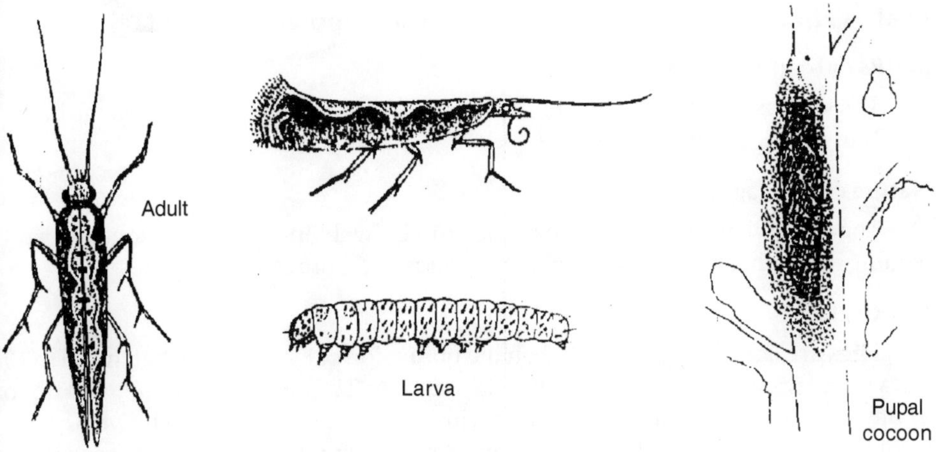

Figure 12.8: Diamond Back Moth

Nature of Damage

Larvae causes serious damage by defoliation. Leaves present a withered appearance or eaten up completely. Larvae damages cabbage, cauliflower. It make holes on them and soiling them with excreta. It is very active in cold season.

Life Cycle

A female moth lays up to 57 eggs. Egg, larval and pupal period lasts for 3-6, 14-21 and 7-11 days, respectively. It pupates in a thin transparent cocoon on the foliage. It completes 8-12 broods in a year.

Management

1. Removal and destroy all debris, remnants and stubbles after harvest of crop and plough fields.
2. Grow 20 days old mustard seedling as intercrop at the time of cabbage or cauliflower planting at 2:25 ratio to attract diamond back moths for oviposition. Periodically spray mustard crop with diclorovos 0.075 per cent at 10 or 15 days interval to avoid dispersal of larvae.
3. Install pheromone traps at 12 per ha.
4. Grow DBM resistant cauliflower variety *i.e.,* Nashik-1.
5. Spray cartap hydrochloride 0.05 per cent or *Bacillus thuringiensis* 2 g per litre or quinalphos 0.05 per cent at primordial stage or head initiation stage when ETL crosses 2 larvae per plant.
6. Spray NSKE 5 per cent along with teepol or sandovit 0.5 ml per litre after primodial stage.
7. Release parasitoids, *Diadegma semiclausum* or *Cotasia plutellae* for effective control.

Leaf Webber, *Crocidolomia binotalis* (Lepidoptera: Pyralidae)

Marks of Identification

The larva is green with a red head and it has longitudinal red strips on the body. It is 2 cm in length.

Nature of Damage

The caterpillar cause damage to the crop by webbing the leaves together and feeding on them. They also feed on flower buds and bore into the pods.

Life Cycle

Eggs are laid in mass of 40-100 on the underside of leaves. They hatch in 5 to 15 days in different conditions. Larval period lasts for 24-27 days in summer and up to 51 days in winter. It pupates in a cocoon within webbed up leaves and flowers or just below ground surface for 14 to 40 days depending on climate. More than one generation may be completed in the season.

Management

1. Remove and destroy the webbed leaves having caterpillars within.
2. Use light trap at 1 per ha to attract and kill adults.
3. Spray carbaryl 0.2 per cent or malathion 0.15 per cent at fortnight interval.
4. Encourage the activity of parasitoid, *Apanteles crocidolomiae*.

Cabbage Borer, *Hellula undalis* (Lepidoptera: Pyralidae)

Marks of Identification

The adult moth is slender, pale yellowish-brown, having grey wavy lines of the forewings. Its hindwings are pale dusky. The larva is creamy yellow with a pinkish tinge and has seven purplish brown longitudinal strips.

Nature of Damage

The caterpillar mine into the leaves, latter on, they feed on the leaf surface, sheltered within the skin passages. As they grow bigger they bore into the heads of cauliflower and cabbage. When attack is heavy, the plants are riddled with worms and outwardly the heads look deformed.

Life Cycle

The female moths lay eggs singly or in clusters on the under surface of the leaves or some other parts of the plant. The eggs hatch in 2-3 days. The caterpillar fed in the head of the cabbage and become full-grown in 7-12 days, after undergoing four moultings. The full grown caterpillar spins a cocoon among the leaves and pupate inside it. The pupal period is about 6 days and the life cycle is completed in 15-25 days.

Management

1. Collection and mechanical destruction of caterpillars in early stage of attack helps to check the infestation.

2. Grow resistant variety of cauliflower *viz.*, Early Patana, ES-96 and ES-97.
3. Spray quinalphos 0.05 per cent or endosulfan 0.1 per cent or malathion 0.1 per cent.

Cabbage Butterfly, *Pieris brassicae* (Lepidoptera: Pieridae)

Marks of Identification

Adult are a snow white butterfly having black marking on wings. The full grown larvae are 40-50 mm long. The young larvae are pale yellow, and become greenish-yellow later on. The head is black. The body is decorated with short hairs.

Nature of Damage

The caterpillar cause damage. The first instar caterpillar just scrape the leaf surface, whereas the subsequent instars eat up leaves from the margins inwards, leaving intact the main veins.

Life Cycle

Eggs are laid in cluster of 50-80 on leaf surface. The eggs are flask shaped. Eggs hatch in 11-17 days. The caterpillar feed gregariously during the early instars and dispers in latter instars. They pass through five stages and are full-fed in 14-22 days. The pupal period lasts for 7-15 days. There are four generations in a year.

Management

1. Collect and destroy caterpillars mechanically in the early stage of attack.
2. Grow resistant/tolerant varieties of cabbage *viz.*, Red rock mammoth, Red rickling, Red drum head.

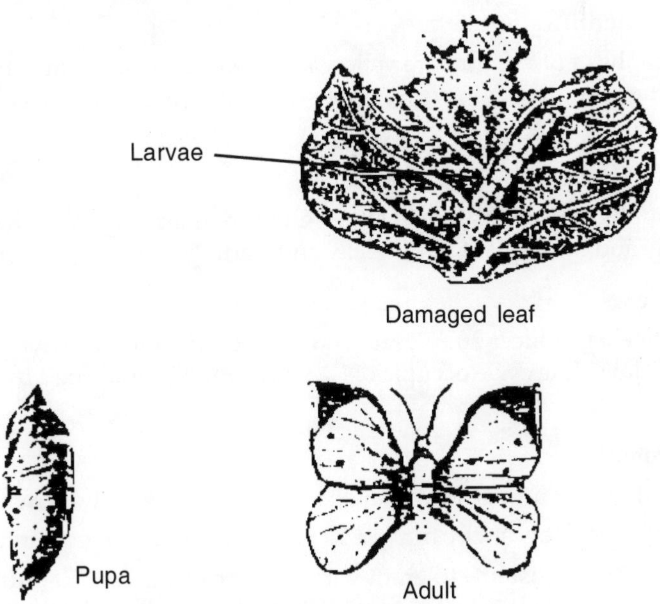

Figure 12.9: Cabbage Butterflies

3. Spray quinalphos 0.05 per cent or endosulfan 0.1 per cent or malathion 0.1 per cent.
4. Encourage the activity of *Apanteles glomeratus*.

Flea Beetle, *Phyllotreta cruciferae* (Coleoptera: Chrysomelidae)

Marks of Identification

Adult beetles are elongate oval in shape and metallic bluish green in colour. Head has impunctate vertex and black antennae.

Nature of Damage

The adults mostly feed on the leaves by making innumerable round holes in the host plants. The affected leaves dry up. A special kind of decaying odour is emitted by the cabbage plants attacked by this pest.

Life Cycle

A single female lay 50-80 eggs in soil. Incubation period last for 5-10 days. The grubs feed on roots and do not cause much damage. Grub period lasts for 9-15 days. Pupal period is of 8-14 days. There are 7-8 generations of this pest in a year.

Management

1. Hand pick and destroy adult beetles.
2. Spray quinalphos 0.05 per cent or endosulfan 0.1 per cent

Aphids, *Brevicoryne brassicae* and *Lipaphis erysimi* (Hemiptera: Aphididae)

Marks of Identification

Nymphs of *B. brassicae* are yellowish green while adults are long and darker in colour than nymphs. *L. erysimi* nymphs and adults are lighter in colour and slightly longer in size.

Nature of Damage

Nymphs and adults infest young shoots and leaves. They produce honeydew profusely and make plant parts sticky and leading to fungal growth on them.

Life Cycle

A female produces 26-67 young ones in her life span. The nymphal period lasts for 11-15 days. Life cycle occupies 11-18 days and there can be 11-20 generations in a year.

Management

1. Remove and destroy damaged plant parts along with nymphs and adults.
2. Use yellow sticky traps at 12 per ha to attract winged aphids.
3. Grow resistant/tolerant varieties of cabbage *i.e.,* Red drum head, Early queen, Red rock mammoth, Glory, Red pickling, Express mail, etc.

4. Spray dimethoate 0.03 per cent or neem oil 2 per cent along with 0.5 ml teepol.
5. Encourage activity of natural enemies: predators, Coccinellids and Syrphids, parasitoids, *Diaeretiella rapae*, pathogens, *Entomopothora corcnata* and *Cephalosporium aphidocola*.

Pea and Beans

Pea Stem Fly, *Ophiomyia phaseoli* (Diptera: Agromyzidae)

Marks of Identification

Adults are metallic black flies, having hyaline wings. Females are slightly bigger than males. Maggots are yellowish in colour.

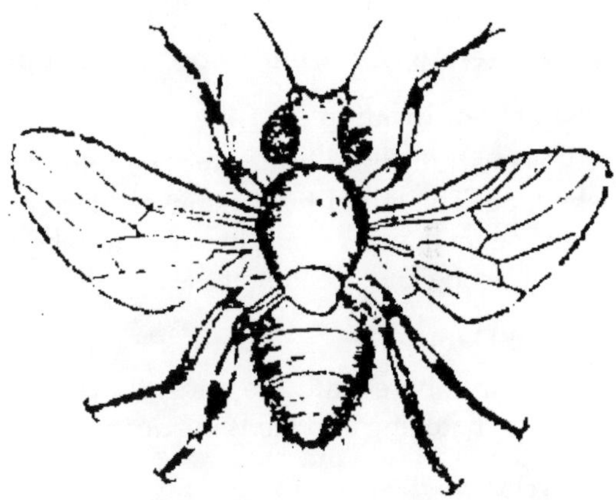

Figure 12.10: Pea Stem Fly

Nature of Damage

Maggots mine the leaves, bore inside the petioles and tender stems and tunnel down wards. Adults also puncture the leaves. Affected leaves turn yellow while the stems drop down and gradually wither away.

Life Cycle

A female lays 14- 64 eggs. Egg hatch in 2-4 days. The maggot mine in the stem. The maggot full fed in 9-12 days. Pupation takes place in the underground portion of affected stem. Pupae are barrel shaped and brown in colour. Pupal period lasts for 18-19 days. There are 8-9 generations in a year.

Management

1. Remove and destroy all affected branches during initial stage of attack.
2. Seed dressing with dimethoate 4 ml per kg of seeds at the time of planting.
3. Spray with 0.03 per cent dimethoate or endosulfan 0.05 per cent.

Spotted Pod Borer, *Maruca testulalis* (Lepidoptera: Pyraustidae)

Marks of Identification

Dark brown with a white cross band in the middle of the forewings and the hind wings are white with a darker border.

Nature of Damage

Presence of semi-solid excreta at the junction of the borehole. Young shoot with dried tip, large scale dropping of flowers. Larva present inside the webbing of leaves, flowers and young pods, faecal material accumulates outside the borehole. It feeds on the seeds by boring into the pods.

Life Cycle

Eggs are elongate oval in shape and light yellow in colour and are laid singly on or near flower buds of host plants. Young caterpillars feed on reproductive parts of flowers and move from one flower to another. Later they web inflorescences with adjacent leaves and developing pods and feed within by boring into the flowers and pods. Full grown caterpillars are light brown in colour with irregular brownish black dorsal, lateral and ventral spots. Incubation period is 2-3 days. Larval stage lasts for 8-14 days and pupal period lasts for 6-9 days.

Blue Butterfly, *Lampides boeticus* (Lepidoptera: Lycaenidae)

Marks of Identification

It is medium sized butterfly. The colour of the wings is violet metallic blue to dusky blue. The tail of hind wings is black and tipped with white. The female is slightly bigger than the male. In males, the abdomen is slender and tapering, while in female it is long and broader at the tip.

Nature of Damage

The larva bores into the buds, flowers and green pods just within couple of hours after hatching and feeds inside the developing grains.

Life Cycle

Eggs are laid on the buds, flowers, green pods and on shoot and leaves. Greenish white in colour, round in shape with a slight depression at the top. Incubation period lasts for 5-7 days. Newly hatched larva is yellowish green in colour with black head and a dark-brown patch on the prothorax and cylindrical body with scattered hair. Full-grown larva is yellowish green to yellowish red sometimes light purple in colour, ventral surface is light green. Whole larva is covered with small setae and marked with irregular black markings. It looks like a slug. Larval stage completed in 10-27 days. Pupa are green in colour later on it darkens and wings are also visible. Pupal development is completed in 10-15 days. The pest breeds through out the year and passes through 5 generations in a year.

Management

1. Grow resistant or tolerant varieties to pod borer.
2. Mechanical destruction of caterpillars in the initial stage of attack.
3. Spray quinalphos 0.05 per cent or endosulfan 0.07 per cent or *Bacillus thuringinsis* 0.1 per cent during 50 per cent flowering stage.

Spiny Pod Borer, *Etiella zinckenella* (Lepidoptera: Phycitidae)

Marks of Identifications

Grayish brown moth, distinct pale-white band along the costal margin of the forewings, hind wings are semi-transparent with a dark marginal line. Orange coloured prothorax.

Nature of Damage

Entrance hole in the green pod disappears and leaves little evidence that the pod is infested. In pods, the larva devours many seeds. The pod always contains a mass of frass and held together by a loosely spun web. Young larva bores into floral parts, making rough and irregular incision.

Life Cycle

Eggs are laid singly or in small groups on immature pods either along the midrib or on the calyx. Freshly laid eggs are glistening white and adhere securely to whatever they touch. Incubation period lasts for 3-5 days. Dorsal surface of mature larva is reddish pink, while the pleural and ventral surfaces of the body are pale-green or creamy-white. Larval period is 12-17 days. Pupa light green in colour changes to light brown or amber. Pupates in the ground at a depth of 2 to 4 cm. pupal period lasts for 14-17 days. Total life cycle occupies 35-36 days.

Management

1. At flower initiation stage spray the crop with endosulfan 35 EC 750 ml per ha and repeat the spray after three weeks.

Bean Aphid, *Aphis craccivora* (Hemiptera: Aphididae)

Marks of Identification

Apterous females are shiny, dark brown or black. Alate forms are greenish black with transparent wings.

Nature of Damage

Colonies of nymphs and adults found on leaves, terminal shoots and pods and suck the plant sap. They are vector of stunt disease in chickpea, rosette of groundnut. Serious pest when the rainfall is low.

Life Cycle

The common mode of reproduction is through vivipary and parthenogenesis, though reproduction by ovipary has been also recorded. A single apterous, parthenogenetic female produced 29 nymphs. The nymphs generally undergo 4 moults before reaching the adult stage. The duration of each instar is usually one day, though in some cases it was even 3 days. Within a day after it becomes an adult the apterous female starts producing its brood. A female reproduced upto a maximum of 12 days.

Management

1. Groundnut intercropped with cereals (*e.g.* millets or maize)reduced the incidence of this pest.
2. Encourage the activities of the predators, *viz., C. septumpunctata, M. sexmaculatus, B. suturalis* and *X. scutellarae.*
3. Spray metasystox (0.1 per cent) or demeton-o-methyl (0.05 per cent).

Potato

Potato Aphid, *Myzus persicae, Aphis gossypii, A. fabae* and *Rhopalosiphum rufiabdominalis* (Hemiptera: Aphididae)

Marks of Identification

Myzus persicae

It is light to dark green or pink with well developed frontal tubercles, which project inwards. The cornicles are long, cylindrical and slightly swollen in the middle. In the winged form, a dark patch can be seen on the abdomen.

Aphis gossypii

Colour variable, ranging from pale yellow to brown or grey black or light to dark green. The tips of the leg joints, eyes and cornicles are black.

A. fabae

Variable in colour, from black to olive green, often with irregular dark pigmented areas over the abdomen. Cornicles slightly black, imbricated, slightly tapering towards exterior. In the winged forms some black bars occur on the abdomen.

Rhopalosiphum rufiabdominalis

Olive green to almost black. Antennae 5 or 6 segmented. Media of forewing sometimes once branched. Usually with reddish blotches at the base of siphunculi.

The first two species *viz., Myzus persicae* and *Aphis gossypii* mainly act as major pest only.

Nature of Damage

Both adults and nymphs are destructive as they suck the sap from the plant. Beside this they act as potential vector of potato leafroll (PLRV) and Y(PLY). They are not serious as pest on the crop but play a vital role in limiting disease form seed production. The losses in yield by the aphid transmitted virus range between 40–85 per cent.

Life Cycle

The winged viviparous females start appearing on potato and other secondary host plants from middle of November onwards. These forms reproduce parthenogenetically and give birth to living young ones. These pass through four nymphal stages. Each stage is of about 1-5 days duration and one generation is completed in about 15 days. This type of asexual reproduction goes on for many generations on the secondary hosy plants.

Management

1. Remove all weeds, hosts susceptible to virus and for aphids especially those having yellow flowers, and volunteer (selfgrown potatoes) plants from within and around the vicinity of field.
2. The haulm cutting of seed crop should be done as soon as the aphid number crosses the critical level *i.e.,*20 aphids per leaves.

3. Application of phorate 10 G @ 1.5 kg a. i. per ha on furrows at planting time will keep the pest under check upto 45-60 days. This should be followed by the need based foliar application of any suitable systemic insecticides such as dimethoate 30 EC @0.03 per cent.

Cut Worm, *Agrotis ipsilon* and *A. segetum* (Lepidoptera: Noctuidae)

Marks of Identification

Adults are medium to heavy bodied moths. *A. ipsilon* moth is dark brown to grayish brown with large areas of black patches on the forewings and thorax. The moths of *A. segetum* slightly smaller than *A. ipsilon*, measures 20 mm in length and 35 mm wings width.

Nature of Damage

Crop damage is caused by the caterpillars only. They feed at night on young shoots or under ground tubers. In the early stages of the crop, the caterpillars cut the stem of the young plants near the ground and feed on the shoots and leaves. After tuberization, they feed by bore and nibble into the tubers affecting both tuber yield and market value.

Management

1. Deep ploughing during hot weather reduce the population of immature stage. A number of birds, such as crow, mynah, starling, etc. feed on the insects that get exposed upon ploughing.
2. Natural enemies play an important role in the management of cutworms. Some of the important parasites of cutworms are, *Broscus punctatus* and *Liogryllus bimaculatus, Auplopus hypsipylae, Ichneumon* sp., *Turanogonia chinensis.*
3. Spray chlorpyriphos 20 EC @ 2.5 litres per ha for spraying the foliage and drenching the ridges on noticing the cutworms attack.

White Grub, *Lachnosterna longipennis* and *L. coracea* (Coleoptera: Scarabaeidae)

Nature of Damage

The damage to potato crop is caused by the grubs in two ways *i.e.*, when they initially feed on rootlets or roots and after tuberization on tubers. The first stage grubs feed on the organic matter available in the soil but they prefer live roots. The damage to potato tubers is caused, mainly by the second and third instar grubs which make large, shallow and circular holes and render them unfit for human consumption.

Life Cycle

The beetles start coming out of the soil at dusk, soon after the first pre-monsoon shower in May end or early June and settle on nearby bushes or trees *viz.*, Acacia, neem, roses, etc., first for mating and afterwards for feeding. The beetles after feeding on the host foliage return to soil (5 to 10 cm deep) early in the morning (before sunrise) for egg laying. The eggs and the first instar grubs are found near the root zones of potato plants between June and August, while second and third instars attack tubers during August–October.

These grubs move down into the soil, with the fall in temperature. These grubs reach upto one meter depth and hibernate in earthen cells. They come up in March-April with the rise in temperature of atmosphere and soil. The full grown grub pupate during April-May for about a fortnight and the beetles emerge between late may and early June.

Management

A. Beetles

1. A majority of beetles are attracted to the light source hence electric or petromax light traps may be operated for mass-collection.
2. Wild shrubs and other hosts of beetles growing in or around the field should be removed, excepting a few which should be sprayed with insecticides before the emergence of beetles.
3. Adults may be killed by spraying the host trees or shrubs with insecticides like chlorpyriphos 20 EC or endosulfan 35 EC or quinalphos 25 EC at 0.1 to 0.2 per cent concentration.

B. Grubs

1. Repeated ploughing before monsoon expose the grubs and pupae for predation by natural enemies such as crows, mynahs, etc., or may be hand collected and destroyed.
2. Flooding of the field, wherever possible for 7-10 days.
3. Apply only well rotten FYM, compost in the field.
4. Apply phorate 10 G or carbofuran 3 G @ 2.5–3.0 kg a. i. per ha in furrows at planting or near plant base at earthing time. Application of granular insecticides at earthing time is more effective.

Potato Tuber Moth, *Phthorimaea operculella* (Lepidoptera: Gelechiidae)

Nature of Damage

This is one of the most destructive pest of potato under the worm and dry environments of field and stores. The larvae attack potato in two ways *i.e.* by mining of younger leaves and feeding on the tubers. Larvae penetrate the leaves and feed within leaf veins or stems of the plant and on tubers in storage depositing the eggs near the eye buds, causing irregular galleries or tunnels deep inside the tuber. Infested tubers are seen at maturity in the field and top of the storage. Pest infestation in field and storage varies from 0-84 and 0-52 per cent in different parts of country.

Life Cycle

There are four stages of this pest namely egg, larva, pupa and adult. On an average, 60-70 eggs are laid on leaves, eye buds of the tubers and soil around the plant. Eggs are laid either singly or in groups of 6 to 7 on an eye bud. The incubation

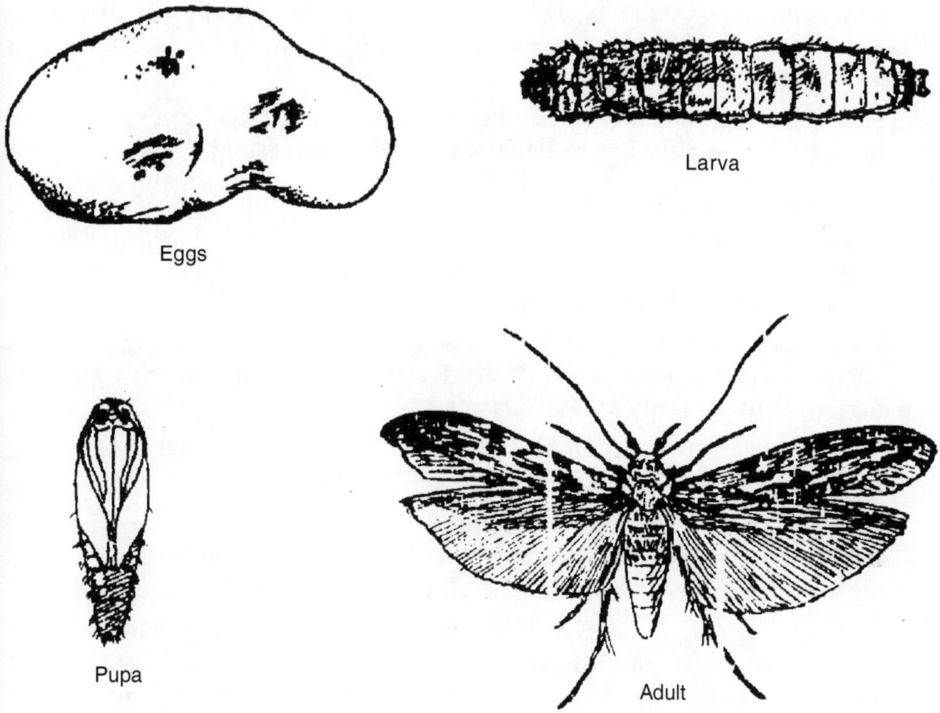

Figure 12.11: Potato Tuber Moth

of eggs, larval and pupal period varies from 2 to 28, 6 to 50 and 6 to 33 days, respectively, depending on the environmental conditions. Life cycle of insect is usually completed within 21-30 days. Several generations are completed throughout the year.

Management

A. Field Control Measures During Crop Growth

Measures to protect the potato crop from planting to harvest are primarily cultural control methods. These are as follow:

Good soil preparation: Adequate soil preparation not only ensures vigorous plant growth but also helps to destroy the resting stages, mostly pupae, of the tuber moth before planting.

Deep planting: Covering tuber seed to a depth of 5-10 cm prevents female moths from ovipositing in seed tubers and keeps larvae from migrating to tubers from infested above ground sprouts. Newly emerged larvae of potato tuber moth (PTM) can burrow to a depth of 10 cm to the seed tubers.

High hilling: High hilling of growing plants protects the developing tubers from ovipositing female and reduces the possibility of larvae reaching the bulking tubers.

Frequent irrigation: Adequate watering and cultivation prevent cracks forming in the soil. Soil cracks allow female moths to reach the potato tubers for oviposition, and provide shelters to adult moths.

Pheromone traps: Commercial pheromones are available for potato tuber moth (PTM). Mass trapping of male moths reduces the probabilities of moth mating, thus causing a drop in egg fertility.

Spray pesticide: Spray monocrotophos 40 EC @ 1.5 litre in 1000 litres of water on 30 days old crop. It may be repeated two weeks before harvesting.

B. Control Measures at Harvest

The two most important control measures at harvest are protecting harvested tubers from ovipositing females and removing crop residues from the field. The following practices should be adopted:

Timely harvesting: During the last phase of the crop *i.e.* tuber filling and plant senescence, the infestation rate accelerates. Delaying harvest by one or two months can increase damage as much as 70-80 per cent

Storing healthy tubers: Only healthy tubers should be stored. Infested tubers should be buried under at least 10 cm soil.

Covering tubers: Female moths become active in the evening and most eggs are laid at that time. Harvested tubers should not remain exposed to ovipositing females overnight. If they cannot stored immediately tubers should at least be covered.

Destruction of harvested residues: Insect pupates in tubers and dry stems left in the field. Moths from these pupae infest the crop. The tubers left in the field become volunteer plants for the pest. For these reasons, all harvest residues must be destroyed.

C. Measures to Avoid Damage in Storage

Cleaning stores: Cleaning floors, walls, and ceilings of stores before storing healthy tubers destroys pupae and other life stages of the moth.

Storing healthy tubers: Tubers should be sorted and infested ones discarded before storing. Tubers exposed to moth oviposition should not be stored, as eggs are commonly overlooked during sorting. Storing infested tubers or those that have been exposed to moth oviposition along with healthy tubers may result in infestation of the entire store with in 3 to 4 months.

Use of repellents: The foliage of some plants, rich in essential oils, such as neem, eucalyptus and *Lantana camara*, repels pest. The leaves are dried under the shade, crushed, and then used to cover the tubers in stores.

Use of pheromone traps in stores: This will reduce the further multiplication of the potato tuber moth and reduce the population (@ 4 traps per 100 cu m).

Microbial pesticides: Use of microbial pesticides *viz., Bacillus thuringiensis* (B.t.) and Granulosis virus (GV), are also very effective in reducing the population of potato tuber moth.

Sweet Potato

Sweet Potato Weevil, *Cylas formicarius* (Coleoptera: Curculionidae)

Marks of Identification

Adult weevils are ant like, slender bodied having elongated snout like bluish

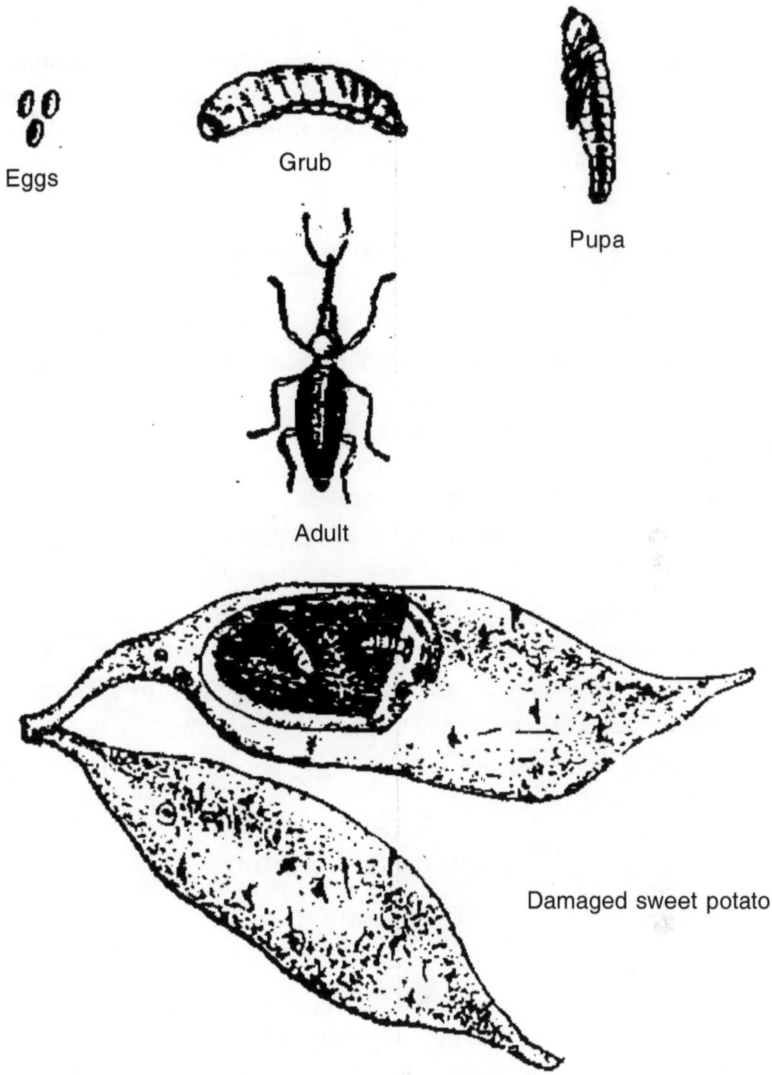

Figure 12.12: Sweet Potato Weevil

brown head with non-geniculate antennae, bright red thorax and legs and brownish red abdomen. Females make small cavities on the tubers or stems and lay eggs singly.

Nature of Damage

It is a specific pest of sweet potato. Damage is caused both grub and adults. The pest infest sweet potato both in field and stores. Grubs bore into stems, because tunneling inside and feed on soft tissues. Grubs and adults bore into tubers both in the field and storage godowns. Affected tubers develop dark patches, which later start rotting. Pest is disseminated from field to field through infested vines and is carried over from season to season by breeding in damaged tubers left in the fields after harvest. The damage caused by this pest is upto 70 per cent.

Life Cycle

Each female lays 100-200 eggs. Grubs are fattish, legless, pale-yellowish white in colour. Pupation takes place in larval burrows. Incubation, grub and pupal stages last for 5-10, 16-20 and 4-8 days, respectively. Life is complited in 4-5 weeks. More then one generations are completed in a year.

Management

1. Remove previous sweet potato crop residues and alternate host, *Ipomoea* sp., and destroy them.
2. Use pest free planting material.
3. Use deep rooting varieties like White Star, CL-44 and Pusa Red to avoid the attack of this pest.
4. Use cut sweet potato tubers (100 g) as trap during 50-80 days after planting (DAP) at 10 days intervals. Set the traps at 5 m apart at 4 PM and collect and destroy adult weevils at 6 AM next day.
5. Dip planting materials in fenthion 0.05 per cent or monocrotophos 0.05 per cent.
6. Rake up soil and earth up at 50 days after planting.
7. Drench soil with endosulfan 0.05 per cent, or spray it if needed.
8. Harvest crop immediately after maturity and destroy the crop residues.
9. Install yellow sticky trap at 12 per ha.
10. In godowns, treat out side of bags containing tubers with malathion 5 per cent.

Onion and Garlic

Onion Thrips, *Thrips tabaci* (Thysanoptera: Thripidae)

Marks of Identification

The adults are slender, yellowish brown and measure about 1 mm in length. The males are wingless whereas females have long, narrow strap-like wings, which are furnished with long hairs along the hind margins. The nymphs resemble the adult in shape but are wingless and slightly smaller in size.

Nature of Damage

Damage is done by adults and nymphs. In onion and garlic the leaves of attacked plants become curled, wrinkled and gradually dry up. The plant do not form bulbs nor do the flowers set seed.

Life Cycle

The pest is active throughout the year. The adult female lives for 2–4 weeks and lays 50-60 eggs singly in slits which are made in leaf tissues with its sharp ovipositor. The egg hatch in 4-9 days. The nymphs pass through four stages and are full fed in 4-6 days. The fully grown nymph pupate in soil at a depth of about 25 mm. The pre pupal and pupal period lasts 1-2 and 2-4 days, respectively. There are 7 generations in a year.

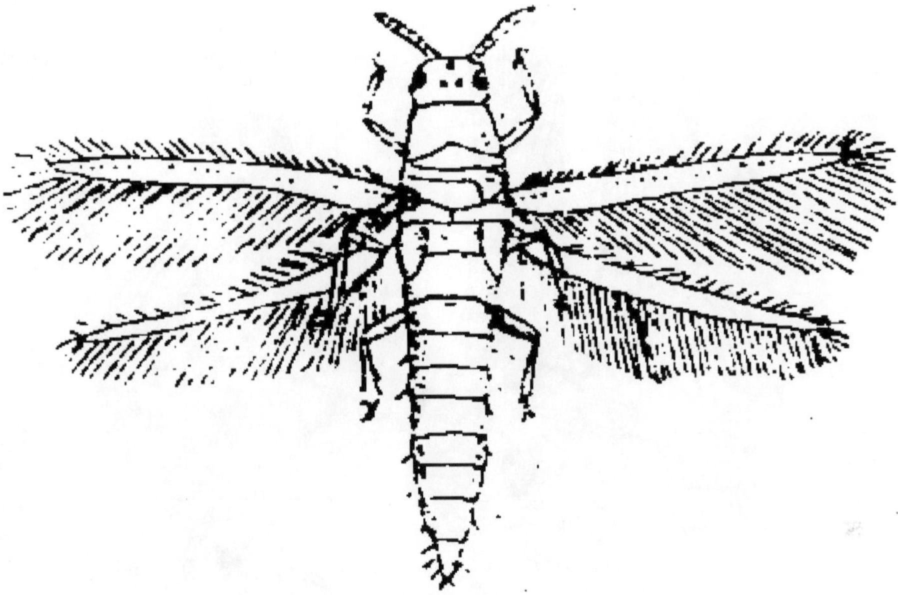

Figure 12.13: Onion Thrips

Management

1. Grow resistant varieties *viz.*, White persian, Grano, Bombay white, Sweet spanish, Crystal wax.
2. Soil application of phorate 10 G @ 2.0 kg a. i. per ha before sowing has been found effective to control onion thrips.
3. Spray 0.05 per cent malathion or 0.075 per cent acephate.

Onion Fly, *Delia (Hylemya) antique* (Diptera: Anthomyiidae)

Marks of Identification

The adult flies are about 6-8 mm in length, slender, large winged, rather bristly. Maggot are small white and about 8 mm long.

Nature of Damage

Only maggots cause damage. Maggots tunnel through bulbs and leaving only the outer sheath. The leaves of the infested plants turn brown from the tip down wards. The attack is not completely destroying the bulbs but it causes subsequently rotting in storage.

Life Cycle

Eggs are laid by female near the base of the plant or cracks or in soil. The incubation period is 2-7 days. After hatching, the maggot mine through bulbs and

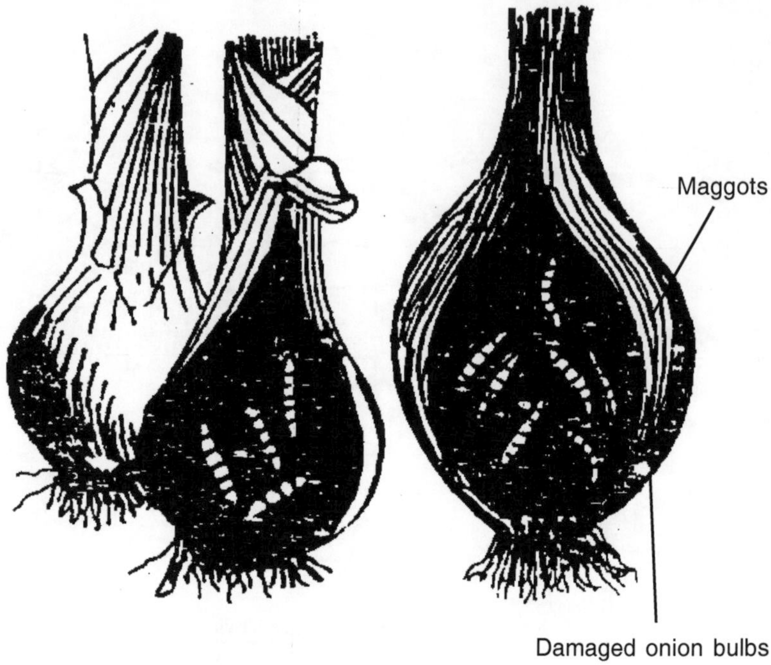

Figure 12.14: Onion Fly

cause damage to bulb. Maggots attain its full growth in 14-21 days and then crawls out the bulb to pupate inside the soil. Pupal period lasts for 14-21 days.

Management

1. Soil application of phorate 10 G at 15 kg per ha or carbofuran 3 G at 25 kg per ha.

Waternut or Singhara

Singhara Beetle, *Galerucella birmanica* (Coleoptera: Chrysomelidae)

Marks of Identification

Adult beetles on emergence are bright yellow but soon become reddish brown. Antennae is brown-black colour. The beetles are 6-8 mm long. Grubs are yellowish-brown. The beetles are sluggish and not active fliers.

Nature of Damage

Grubs and adults feed usually on leaves and sometime also on petioles and integument of the fruits. Full grown grubs are oracious feeders and more destructive than adult beetles.

Life Cycle

A single female lays on an average 101 eggs in clusters of 5-8 eggs, glued firmly to the upper surface of leaves. Pre-mating period is about 2 days, during July and

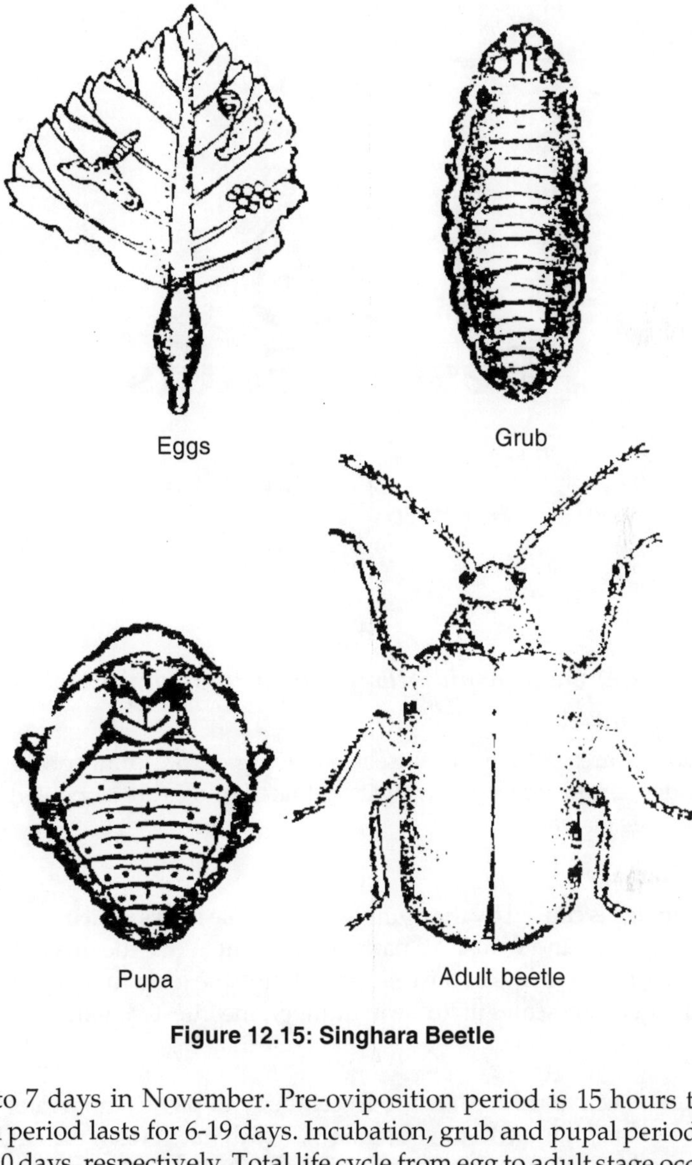

Figure 12.15: Singhara Beetle

extends upto 7 days in November. Pre-oviposition period is 15 hours to 4 days. Oviposition period lasts for 6-19 days. Incubation, grub and pupal periods are 4-9, 9-20 and 5-10 days, respectively. Total life cycle from egg to adult stage occupies 18-32 days. Longevity of male and females is 11-28 and 13-33 days, respectively. The entire life cycle of the beetle and its immature stages is passed on the leaves of water nut plant. The pest is active during August- September.

Management

1. Mechanical collection of adults and egg-masses from the leaves and their destruction in early stages of crop is quite effective.
2. The pest can be controlled by dusting malathion 2 per cent @ 25 kg per ha.

13
Pest of Fruits

Apple

San Jose Scale, *Quadraspidiotus perniciosus* (Hemiptera: Diaspidae)

Marks of Identification

The adults are grayish tiny insects showing sexual dimorphism, the female being rounded, 2 mm across and wingless while the males are elongated and winged, bearing 2 wings instead of four.

Nature of Damage

The damage is caused by the nymphs and female scales which suck the sap from the twigs, branches and fruits. All parts of the plant above the ground are attacked and the injury is due to loss of the cell-sap. At first, the growth of the infested plants is checked, but as the scales increase in number, the infested plants may die.

Life Cycle

The pest is active from March to December and passes the winter in nymphal stage. The female gives birth to young ones, which hatch from the eggs developed within her body. Each female may give birth to 200-400 nymphs. They become full grown in 3-40 days and the females again start giving birth to young ones within the next 50-53 days. There are four to seven overlapping generations in a year.

Management

1. For effective management of this pest orchard sanitation should be given priority. Infested pruned material should be collected immediately and burn.
2. The parasite, *Encarsia perniciosa* may be released to check the overwintering population on wild host plants growing around.

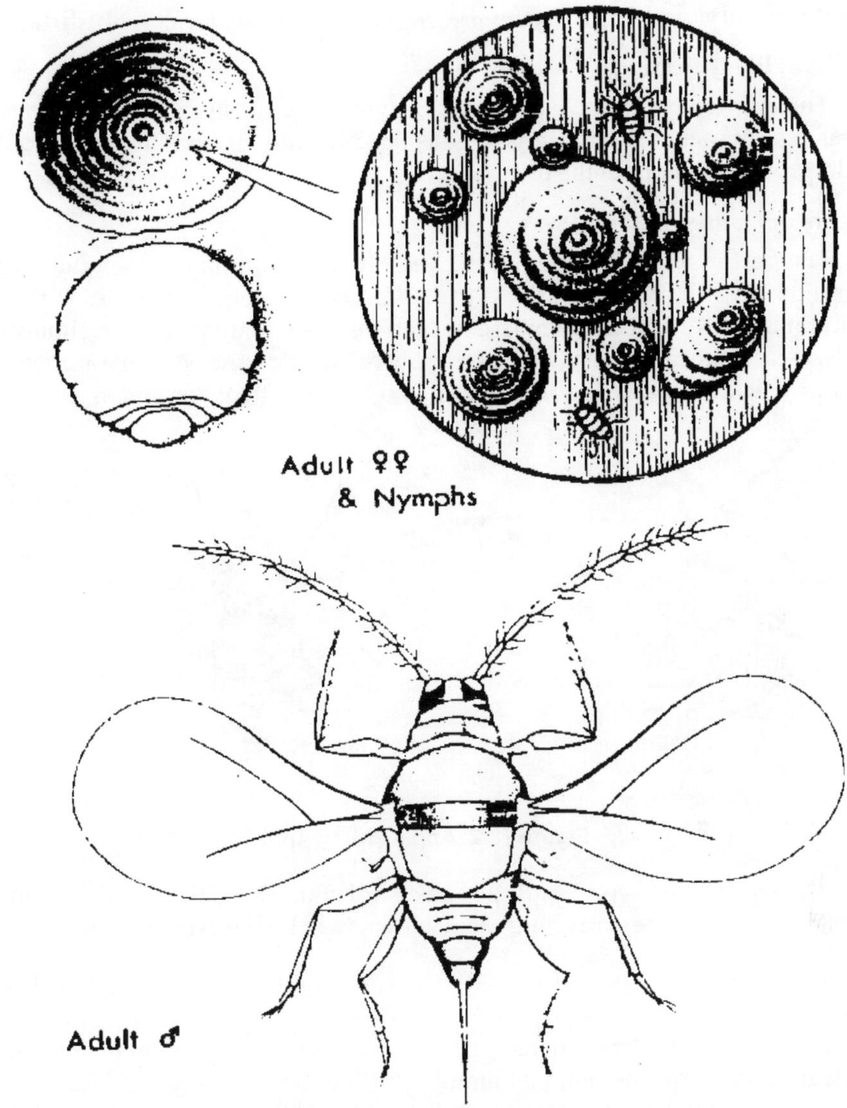

Figure 13.1: San Jose Scale

3. Spray diesel oil emulsion + Bordeaux mixture is to be emulsified and diluted 5-6 times before spraying or spray 7.5 litre of ESSO tree spray oil emulsion in 250 litres of water per ha during the winter season when the trees are in dormant stage and completely defoliated.
4. In case of severe infestation spray methyl demeton 25 EC (1.25 litre per ha).
5. To protect the plants in the nursery apply carbofuron granules 0.75–1.0 g a. i. per plant.

Apple Wooly Aphid, *Eriosoma lanigerum* (Hemiptera: Aphididae)

Marks of Identification

The infested plants have pale green leaves and whitish cottony patches on the stems and branches. Characteristic galls or knots are formed on roots and other underground portions of the plants.

Nature of Damage

The pest is active throughout the year. It attacks primarily the underground roots but winged forms also attacks trunks, branches, stems, twigs, leaf petiols and fruit stalks. Upward and downward migrations are accentuated during hottest and coldest season, respectively. Due to desaping caused by these pests, the affected trees present a sickly appearance, lose vigour and the growth of these trees as also their

Figure 13.2: Apple Wooly Aphid

fruiting capacity are adversely affected, in case of young tree, the roots disintegrate to such an extent that these trees are easily blown over by the even moderately strong winds.

Life Cycle

The pest overwinters either as egg or young nymph on the roots of the host tree. The eggs hatch and the nymphs mature during spring. There are four nymphal instars and the total duration of nymphal stage lasts for 32-43 days. There may be 13 generations in a year.

Management

1. To prevent damage by this pest, use resistant root-stock like Golden delicious, Northern spy and Morton stocks 778, 779 and 793.
2. Release exotic parasitoid, *Aphelinus mali.*
3. Select healthy plant material from nursery and then before planting in the orchard, treat them with chlorpyriphos 0.05 per cent.
4. Spray 500 ml of nicotine sulphate 40 EC or 800 ml of malathion in 500 litre of water per ha.

Codling Moth, *Cydia pomonella* (Lepidoptera: Tortricidae)

Marks of Identification

The adult moth is small, about 12-14 mm in wing span and is 6-8 mm long. The forewings are dark grayish and are marked with wavy lines and a copper coloured metallic eye like circle towards the outer margin. The hindwings are pale grey. The full grown larvae are 16-22 mm long and are pinkish or creamy-white in colour with a brown head. The larvae have eight pairs of legs.

Nature of Damage

Damage is caused by larvae which burrow into the fruit and feed on the pulp. The infested fruits cannot be marketed for human purpose.

Life Cycle

A single female may lay about 100 eggs in her life time. The egg hatch in 4-12 days. The young larvae after emergence enter into the fruit through the calyx. The larval period lasts for 3-4 weeks. At this stage the larva burrows its way out of the apple fruit and fall s to the ground and spins a silken cocoon in which it transforms into a yellowish-brown pupa. The pupal period is completed in 8-14 days. There are two generations in a year.

Management

1. The orchard should be kept clean of all the debris and weeds and remove loose barks from the old trees to prevent the hibernating larvae to find shelter.
2. Pluck and collect fallen infested fruits to burry or burn them to destroy the hiding larvae.
3. Enforce quarantine rules strictly to prevent spread of this dangerous pest.
4. Spray endosulfan 0.07 per cent to protect the fruits from this pest.

Gypsy Moth, *Lymantria obfuscate* (Lepidoptera: Lymantriidae)

Marks of Identification

The female moths are dark grey and they have atrophied wings. The males are comparatively more active in moving around and in mating. Female moths are apterous type, are therefore unable to fly. The caterpillars are 40–50 mm long and is clothed in tufts of hairs.

Nature of Damage

The caterpillars are gregarious and they eat voraciously at night time. Their habit to defoliate the host trees completely results in the failure of fruit formation.

Life Cycle

The eggs are laid under the loose bark and covered over with yellowish-brown hairs. The eggs over-winter as such and hatch in March-April. The larval period completed in 66-100 days. The pupal formation takes place in soil among debris and pupal stage lasts 9-21 days. The male moth lives for 4-10 days and the females for 11-31 days. One generation is completed in a year.

Management

1. Collect and destroy sluggish females and the egg-masses that are so conspicuous and easy to locate.
2. Spray endosulfan 0.07 per cent to protect the fruits from this pest.

Mango

Mango Hopper, *Amritodus atkinsoni* and *Idioscopus clypealis* (Hemiptera: Cicadellidae)

Marks of Identification

Freshly hatched nymphs are wedge shaped and whitish in colour with two small red eyes, gradually with each moulting the colour changes to yellow, yellowish-green, green and ultimately to greenish-brown. Adults are also wedge shaped having greenish brown body and pale yellow vertex. Forewings are thicker than hind wing, bronzy sub-hyaline with veins pale yellow and a white line along the costal margin forming distinct mid longitudinal line when the insect is at rest. Adult of *A. atkinsoni* are darker in colour and bigger in size, male being 4.2 to 4.8 mm long and females 4.7 to 5.1 mm those of *I. clypealis* are smaller and narrower, males being 3.4 to 3.7 mm long and females 3.6 to 3.9 mm.

Nature of Damage

Enormous number of nymphs is found clustering on the inflorescences and sucking sap. The infested flowers shrivel, turn brown and ultimately fall off. On

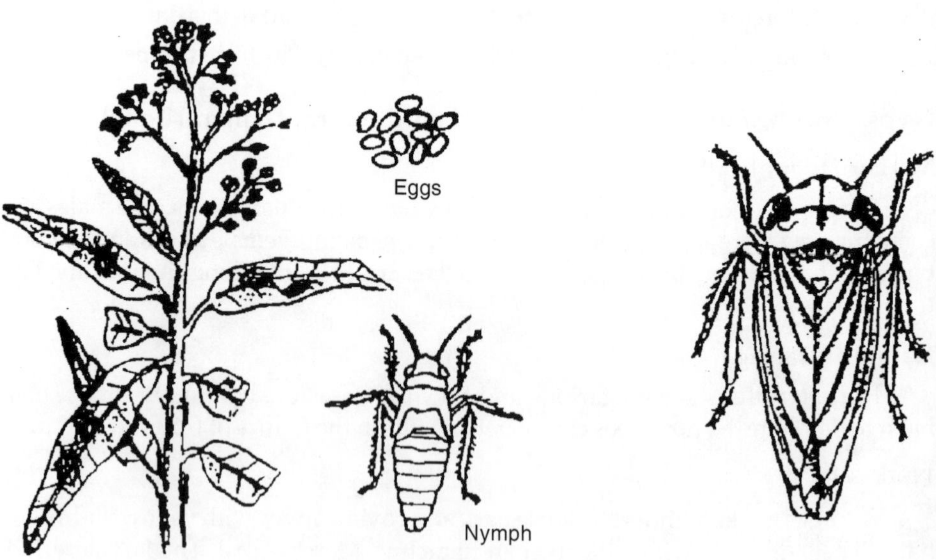

Figure 13.3: Mango Hopper

attaining maturity the hoppers leave the blossoms and move on leaves and trunks of the trees. Swarms of adults are commonly seen hovering in mango groves and sitting on all plant parts. Both nymphs and adults suck the sap usually from the ventral surface of leaves. As a result, growth of the trees is stunted. The hoppers also excrete honey dew which encourages the development of fungi, resulting in growth of sooty moulds on dorsal surface of leaves, branches and even fruits.

Life Cycle

The pest is active throughout the year but during the hot months of May-June and the cold months of October-November, only the adults are found sitting in thousands on the bark of trunks and branches. A female deposits 100 to 200 eggs. The eggs hatch within 4-7 days. The nymphs become full grown in three stages, in 8-13 days. The life cycle from the time eggs laid to the time the adult appear, takes 15-19 days.

Management

1. High density planting should be avoided as it provides favourable condition for hopper multiplication.
2. Regular irrigation should be given in mango orchard to prevent intermittent flushes.
3. Avoid waterlogging or damp conditions.
4. In case of dense orchards, prune some of the branches during early winter to have better light interception.
5. Spray endosulfan 0.07 per cent or phosalone 0.075 per cent or monocrotophos 0.04 per cent or decamethrin 0.0015 per cent. Above insecticides should be sprayed alternatively whereas the hopper population exceeds the count of 5 nymphs per inflorescence.

Stem Borer, *Batocera rufomaculata* (Coleoptera: Cerambycidae)

Marks of Identification

The adults are longicorn beetles, well built, large and pale grayish, measuring about 5 cm in length and 2 cm in breadth. The beetle is provided with long legs and antennae and a dirty white band, extending from head to tip of the body on each side. A number of dirty yellowish spots are present on the elytra. Head is distinct with large prominent eyes and the pronotum is ornamented with two crescent orange yellow spot. Full grown larva is a stout yellowish or white fleshy grub, measuring about 6 cm in length. Its head is dark with strongly developed mandibles.

Nature of Damage

This pest is quite serious when it appears since it attacks the main stem and the branches of the trees. When the stem in any branch is attacked the sap and masses of frass exude from the bored hole. Often the damage may be visible by falling off the leaves of the attacked branches and sudden collapse of the branches. In case of severe damage the entire tree is often killed. The damage is done either to the roots or the

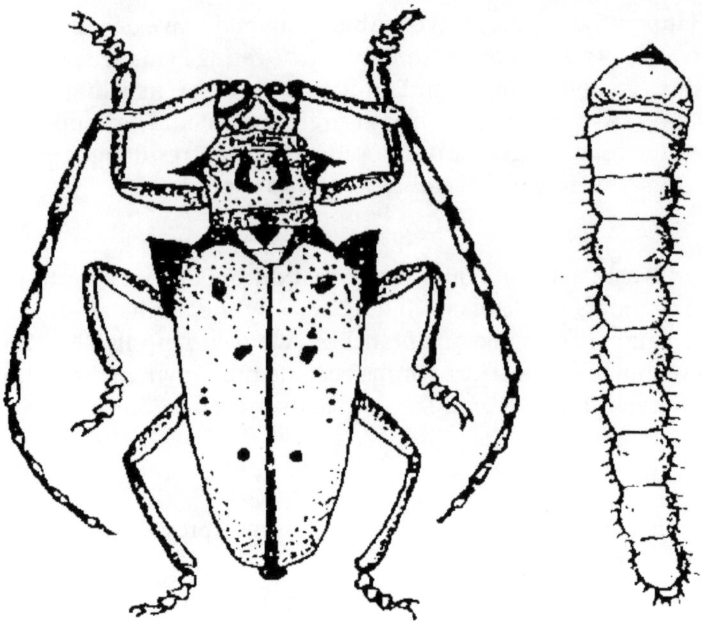

Figure 13.4: Stem Borer

stems. As the grub feds, a harmonious sound is produced by its mandibles. On this account, the insect is popularly known as violin beetle.

Life Cycle

Copulation generally commences 2 days after the adult emerged. The pre-oviposition and oviposition period are on an average 1-1.5 and 22 days, respectively. Female lays eggs parallel to the bark layers in depth of 2.5 to 4.0 mm. Eggs are laid mostly on trunks and primary branches. The incubation period is 7-13 days. Larval period is 140-160 days. The pupal period varies from 19-36 days. Adult live for 25-26 days. The total life cycle completed in 170–190 days.

Management

1. Cut and destroy the infested branches with grubs and pupae within.
2. Beetles wherever found in the garden should be collected and destroyed.
3. The grub should be extruded, through hooked wires or destroy by injecting dilute kerosene or petrol or 0.02 per cent dichlorvos @ 5 ml per hole and then the holes be sealed with the mud.
4. To kill effectively the stem borer larva hidden in up-down tunnel through insecticide, apply the insecticide solution with the help of syring (10 ml).

Fruit Fly, *Bactrocera dorsalis, B. zonata* (Diptera: Tephritidae)

Marks of Identification

The adult fly is stout and measures 14 mm across the wings and 5 mm in maximum length. The flies are strong fliers and can fly up to two kilometers in search

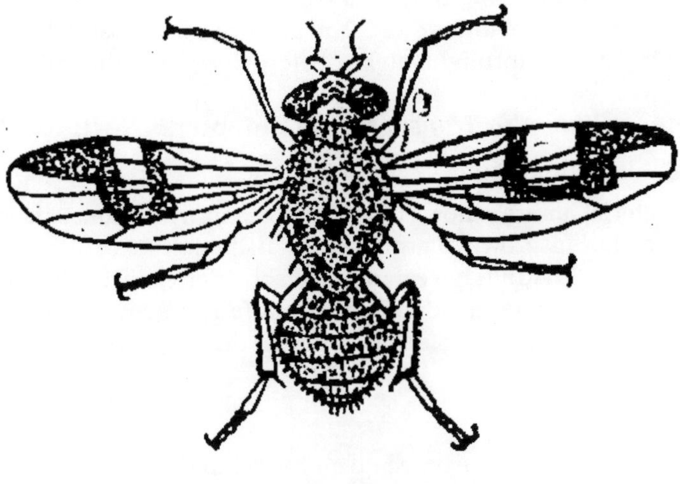

Figure 13.5: Mango Fruit Fly

of food. The fly is brown or dark brown in colour with hyaline wings and yellow legs. The thorax is ferruginous without yellow middle stripe. The abdomen is conical in shape and dark brown in colour. The young maggot is white and translucent.

Nature of Damage

The dark puncture caused by the oviposition of adult fly is not very conspicuous as its colour blends with the dark green colour of the fruit. It is very clearly visible in some yellow and pale brown varieties. The maggot on hatching feed on the pulp of the fruit for few a days and a brown rotten parch appears on the fruit surface. The mesocarp becomes dirty brown. Infested fruits finally fall on the ground. The fruit is affected from late April to June.

Life Cycle

A female lays, on an average 50 eggs in one month. Eggs hatch in 3-10 days. As maggots develop, they pass through 3 stages in the ripening pulp and are full-grown in 6-29 days. They leave fruit and move away by jumping in little hope. On reaching a suitable place, they burry themselves into soil and pupate. In 6-44 days, they emerge as flies and reach ripe fruit for further multiplication. Life cycle is completed in 2-13 weeks and many generations *i.e.*, 10-12 are completed in a year. Flies are present in the field all through the year.

Management

1. Collect and destroy fallen and infested fruits by dumping in a pit and covering with a thick layer of soil.
2. Use parasitoids *Opius compensatus* and *Spalangia philippinesis*.
3. Use methyl eugenol lure trap (25 per ha) to monitor and kill adults or prepare methyl eugenol and malathion 50 EC mixture at 1:1 ratio and take 10 ml mixture per trap.

4. Use bait spray combining molasses or jaggery 10 g per litre and one of the insecticides, malathion 50 EC 2 ml per litre, dimethoate 30 EC 1 ml per litre two rounds at fortnight interval before ripening of the fruit.

Shoot Borer, *Chlumetia transversa* (Lepidoptera: Noctuidae)

Marks of Identification

Young caterpillars are yellowish-orange in colour with characteristic dark brown prothoracic shield. Full grown caterpillars are dark pink with dirty spots and measure 20 to 24 mm in length. Adults have thorax and abdomen clothed with rufous, fuscous and gray scales. Forewings are dark gray beautifully patterned with wavy design. Hindwings are fuscous, apical side being darker than proximal side. Wing expanse is 15-20 mm.

Nature of Damage

The newly hatched larvae bore into midribs of the leaves and feed therein for 2-3 days, and thereafter, they come out and bore into the tender shoots. They make tunnels downwards up to 100 mm to 150 mm in length and expelled out excreta through the entrance hole and the shoot becomes hollow. The affected shoots show dropping of leaves and give a wilting look. The attack is noticed during period when there is new flush on the trees and saplings. The young caterpillars ate attacked during the earlier part of April with the commencement of hot winds. The larvae also eats young leaves and inflorescence.

Life Cycle

The adult female lays eggs singly on the tender leaves and hatch within 2-3 days. The caterpillar takes 10-12 days to mature and then it leave the tunnel and enters into the cracks, and crevices of bark of the tree, dried malformed panicles and also in the soil for pupation. It makes a silken cocoon and within that it pupates. The

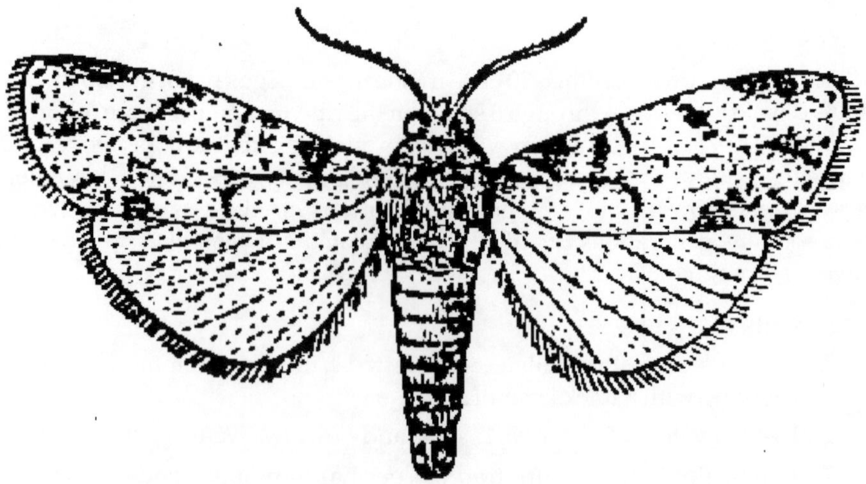

Figure 13.6: Shoot Borer

pupae are unaffected till the onset of monsoon. Generally from these pupae the moths emerges out in 15-18 days. The total life cycle completed in one and half month's period.

Management

1. To control this pest, clip off and destroy promptly affected shoots in the initial stage of attack.
2. Spray carbaryl 0.2 per cent or endosulfan 0.07 per cent or monocrotophos 0.04 per cent 2-3 times at three weeks interval, commencing from initiation of new flush of leaves.

Leaf Gall Midge, *Procontarinia matteiana* (Diptera: Cecidomyiidae)

Marks of Identification

The adult is a minute, yellowish colour with grayish black. The males are larger than the females. The head bears two large compound eyes and long antennae. There are two tiny transparent wings. The leg are slender and long each bearing strongly arched claw with additional dent.

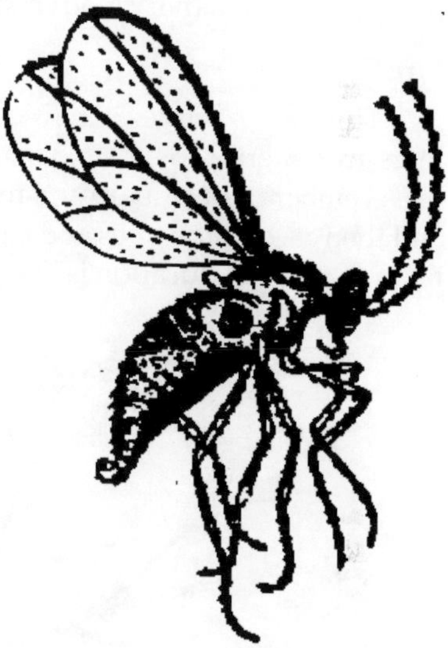

Figure 13.7: Leaf Gall Midge

Nature of Damage

The pest lays its eggs on ventral surface of leaves, on hatching the maggot bore inside the leaf tissues and feed within, resulting in formation of small raised wart-like galls on the leaves. The affected leaves get badly deformed and drop prematurely.

Life Cycle

The eggs are laid in tender leaves and life embedded within leaf tissue. The oviposition sites are marked by small reddish spots. Within a week after oviposition, the individual galls having eggs show symptoms as lenticular thickenings. The incubation period lasts for 3-4 days. The larval period lasts for 2-12 months. There are three overlapping generations in a year. The adults live for about two days.

Management

1. Cultural practice such as summer ploughing has been found useful in reducing the midge population in the following years.
2. Cut the heavily infested leaves at the early stage of infestation and destroy them.
3. Foliar application of monocrotophos 0.04 per cent or dimethoate 0.03 per cent or methyl-o-demeton 0.03 per cent should be applied at 15 days interval during new flush of leaves.

Mealybug, *Drosicha mangiferae* (Hemiptera: Margarodidae)

Marks of Identification

This pest is univoltine and commonly called as giant mealybug. The nymphs and adult females are flat, oval, waxy whitish insects, sometimes mistaken for fungal growth. Adult females are wingless while males are crimson coloured bugs with two dark brownish black wings and cause no damage, except fertilizing the female.

Nature of Damage

Nymphs and adult females suck the sap from inflorescence, tender leaves, shoots and fruits. Excessive desaping of infested tissues leads to their wilting, drying and ultimately fruit setting is affected. They secrete honeydew over which sooty mould (*Capnodium mangiferae*) develops, as a result of which leaves and inflorescence become shining black and sticky. Due to the growth of sooty mould on the leaves photosynthetic activity is adversely affected.

Life Cycle

This pest is active from December to May and spends rest the year in egg stage. Eggs are generally deposited in ovisacs in soil under the trees. They hatch at the end of December or in January and nymphs feed for a month on low-growing vegetation and then ascend the trees and clusters on the growing shoots and suck the sap from them. Nymphal duration is 76- 134 days. Adult duration is 22-27 days.

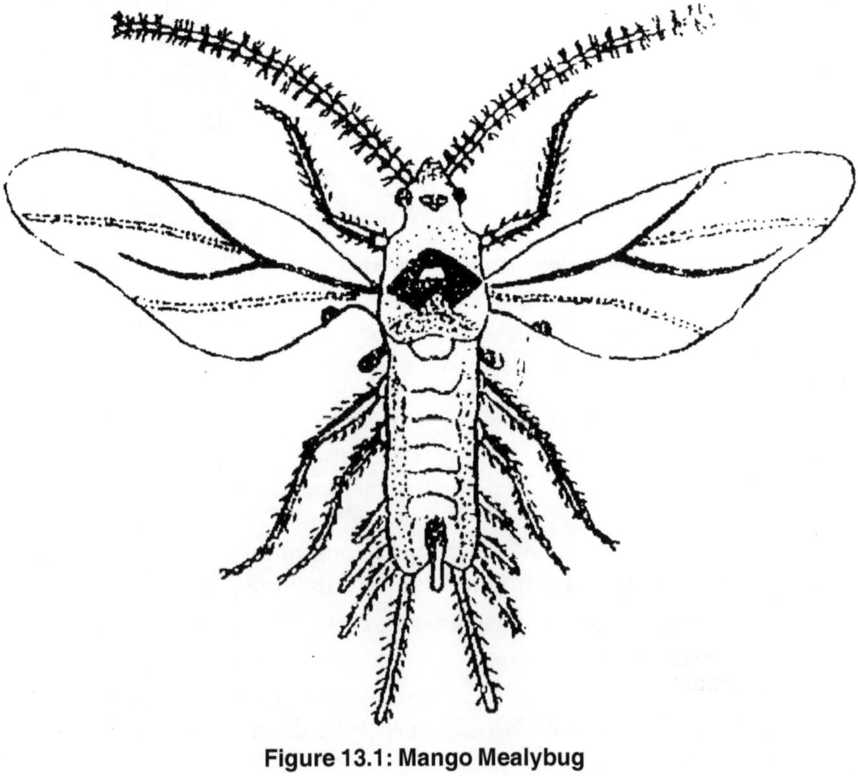

Figure 13.1: Mango Mealybug

Management

1. Remove weeds from orchards which act as additional hosts for mealybug.
2. Ploughing or digging of the soil at depth of 15 cm around the tree during summer exposes eggs to natural enemies and extreme sun heat.
3. Rake the soil around the tree trunk and mix methyl parathion 2 per cent dust @ 200 g per tree.
4. Nymphs should be prevented from crawling up the trees by applying 400 gauge alkethene or plastic sheet 25 cm wide around the trunk about one meter above the ground level. Before tying the band, paste a layer of mud or wet soil around the tree trunk, to make that portion smooth and then wrap the alkethene or plastic sheet and tie it on both sides with sunhemp thread.
5. Nymphs found congregating below the lower edge alkethene bend should be killed mechanically or by spray methyl parathion 0.05 per cent.
6. Foliar application of methyl parathion 0.075 per cent or monocrotophos 0.05 per cent or diazinon 0.1 per cent should be applied as and when the pest appears on mango trees.

Mango Stone Weevil, *Sternochetus mangiferae* (Coleoptera: Curculionidae)

Marks of Identification

It is a short stoutly built, ovoid, dark-brown weevil and found inside the stone of fruit or in its pulp.

Nature of Damage

The insect attacks mango varieties with a relatively soft flesh. The injury caused by the grub feeding in pulp sometimes heals over but a certain number of fruits always get spoiled when the weevils make an exit through ripe or near ripe mango fruits.

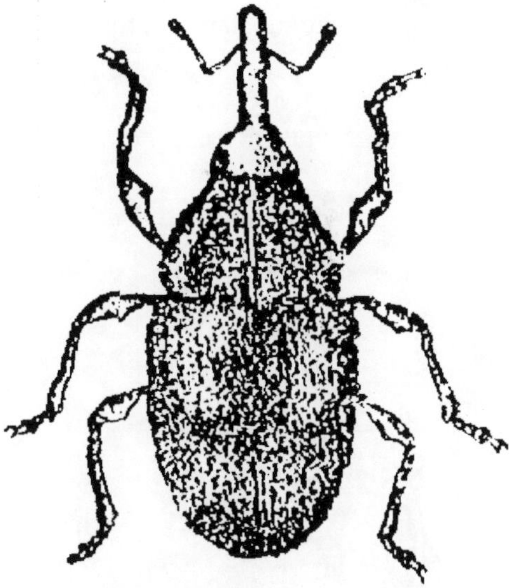

Figure 13.9: Mango Stone Weevil

Life Cycle

The weevil lay eggs in the skin or ripening fruit. On emergence from the egg, the grub further inwards, eating its way through the unripe tissue until it bores into the embryo of the mango-stone. It pupates inside the stone and the weevil cuts its way through the stone. The life cycle is completed in 40-55 days and there is only one generation in a year.

Management

1. Field sanitation including collection and destruction of all fallen fruits at weekly interval will destroy adults.

2. Regular fallen fruit sampling, starting from pea size of fruit, can help in forecast of the pest and helps in timing of first spray.
3. Racking of soil below the tree in October-November and March.
4. At pea to marble size, spraying of fenthion 0.08 per cent followed by monocrotophos 0.05 per cent after two weeks check the pest population.

Citrus

Citrus Psylla, *Diphorina citri* (Hemiptera: Aphalaridae)

Marks of Identification

The adult is small 3-4 mm in length, active in habits, and rest on the leaf surface with closed wings, the tail end of the body being turned upward. The insect is brown with its head lighter brown and pointed. The wings are membranous, semi-transparent with a brown bend in the apical half of the forewings. The hindwings are shorter and thinner than the forewings. The nymphs are flat, louse like and orange yellow creatures, and are seen congregated in large numbers on young leaves and buds.

Nature of Damage

Both nymphs and adults suck the plant sap but the nymphs are more destructive to the plants, they suck the cell sap from tender shoot, leaves and flowers causing curling of leaves, defoliation and drying of twigs. It is also reported that the insect produces a toxic substance in the plant as a result of which the fruit remain under

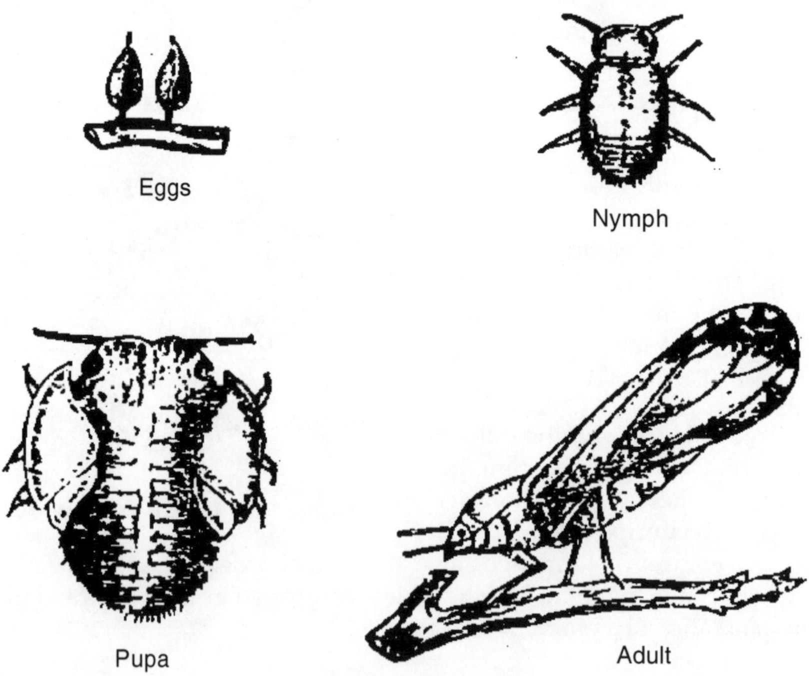

Figure 13.10: Citrus Psylla

sized and poor in juice. The insect is also responsible for spreading the greening virus.

Life Cycle

A single female lay about 800 eggs on tender leaves and shoots after the pre-oviposition period of about 24 hours. The eggs are laid either singly or in groups of two or three. The egg hatch in 10-20 days in winter and 4-6 days in summer. There are five nymphal stages and the development is completed in 10-21 days. There are great variations in the longevity of adults in different seasons. There are 8-9 overlapping generations in a year.

Management

1. Spray suspension of neem or *datura* (*Ipomea fistullosa*) leaves extract (10 per cent) or azadirachtin 0.03 per cent or neem oil 0.5 per cent to control the pest.
2. Spray methyl-o-demeton 0.025 per cent or dimethoate 0.03 per cent or malathion 0.05 per cent or monocrotophos 0.04 per cent or phosalone 0.07 per cent or endosulfan 0.07 per cent at peak activity of this pest.
3. The optimum dose of fertilizers should be applied in citrus orchards as higher NPK dose fertilizer helps higher infestation of psylla than lower NPK dose of fertilizer.
4. Trees should be planted in wider spacing (6.0m x 6.0 m and 6.0 m x 4.5 m) to minimize the infestation of this pest.

Leaf Miner, *Phyllocnistis citrella* (Lepidoptera: Phyllocnistidae)

Marks of Identification

The adult is a tiny moth measuring 4.2 mm across the wings. On the front wings there are brown strips and prominent black spots along the tips. The hind wings are pure white and both pairs are fringed with hairs. The full grown larva measures 5.1 mm in length, apodus and is pale yellow or pale green with light-brown well developed mandibles.

Nature of Damage

The caterpillar attack only young and tender leaves, old leaves are avoided. The larvae make serpentine mines in the leaves feeding on epidermal cells of the leaf leaving behind the remaining leaf tissue quite intact. The mining larvae feed actually more on sap than on solid tissue of the leaf. The mined leaves turn pale, set destroyed and may dry up. The attacked leaves remain on the plant for a considerable long time and the damage gradually spreads to fresh leaves. Heavily attacked plants can be spotted from a distance and young nurseries are most severely affected, the young plants of orange and grape fruits may not even survive. In large trees, the photosynthesis is adversely affected, vitality is reduced and there is an appreciable reduction in yield. Attack by this pest also encourages the development of citrus canker.

Life Cycle

The pest is active throughout the year and breeds on young growth. The duration of various stage depends upon the prevalent temperature. The moths lay about 83-127 eggs singly in 3 days on young leaves or tender shoots, usually on the lower surface, particularly near the midrib. The eggs hatch in 2–10 days. Larval duration is 5-30 days. They settle down in enlargements of the galleries near the leaf margin, and spin cocoons for pupation. The pupal stage lasts for 5-25 days. The life cycle is completed in 12-55 days and about 16 overlapping generations are produced in a year.

Management

1. Prune the heavily affected parts during winter and burn the same for effective control.
2. Spray dimethoate 0.03 per cent on need base (one larva per 15 cm twig length) for effective control of the pest.
3. Spray suspension of neem or datura (*Ipomea fistullosa*) leaves extract (10 per cent) or neem seed extract 2 per cent for the control of the pest.
4. Spray methyl-o-demeton 0.025 per cent or dimethoate 0.03 per cent or monocrotophos 0.04 per cent at peak activity (May-June) of the pest.
5. The varieties, Adinima lime and Zumukhiya of acid lime are found tolerant against the attack of leaf miner.
6. Spray nursery plants 5 ml of fenvalerate 20 EC or 10 ml of cypermethrin 10 EC or 35 ml of decamethrin 2.8 EC or 15 ml of monocrotophos 36 EC in 10 litre of water at fortnightly intervals.

Lemon Butterfly, *Papilo demoleus* (Lepidoptera: Papilionidae)

Marks of Identification

The full grown caterpillar is yellowish green, has a hornlike structure on the dorsal side of the last body segment and is 40 mm long and 6.5 mm wide. The black and white markings make the larvae look like bird dropping. The adult is a large beautiful butterfly,28 mm in length and 94 mm in wing expanse. Its head and thorax are black, being a creamy yellow coloration on the underside of the abdomen. Its wings are dull black, ornamented with yellow markings. The general coloration on the underside of the wings is slightly paler and the markings are also larger. The antennae are black and have club like structures at their ends.

Nature of Damage

The young larvae feed only on fresh leaves and terminal shoots. Habitually, they feed from the margin inwards to the midrib. In later stage, they feed even on mature leaves and sometimes the entire plant may be defoliated. The pest is devastating in nurseries and its damage to foliage seems to synchronize with fresh growth of citrus plants. Heavily attacked plants bear no fruits.

Life Cycle

The female butterfly lays eggs on tender shoots and fresh leaves, mostly on the

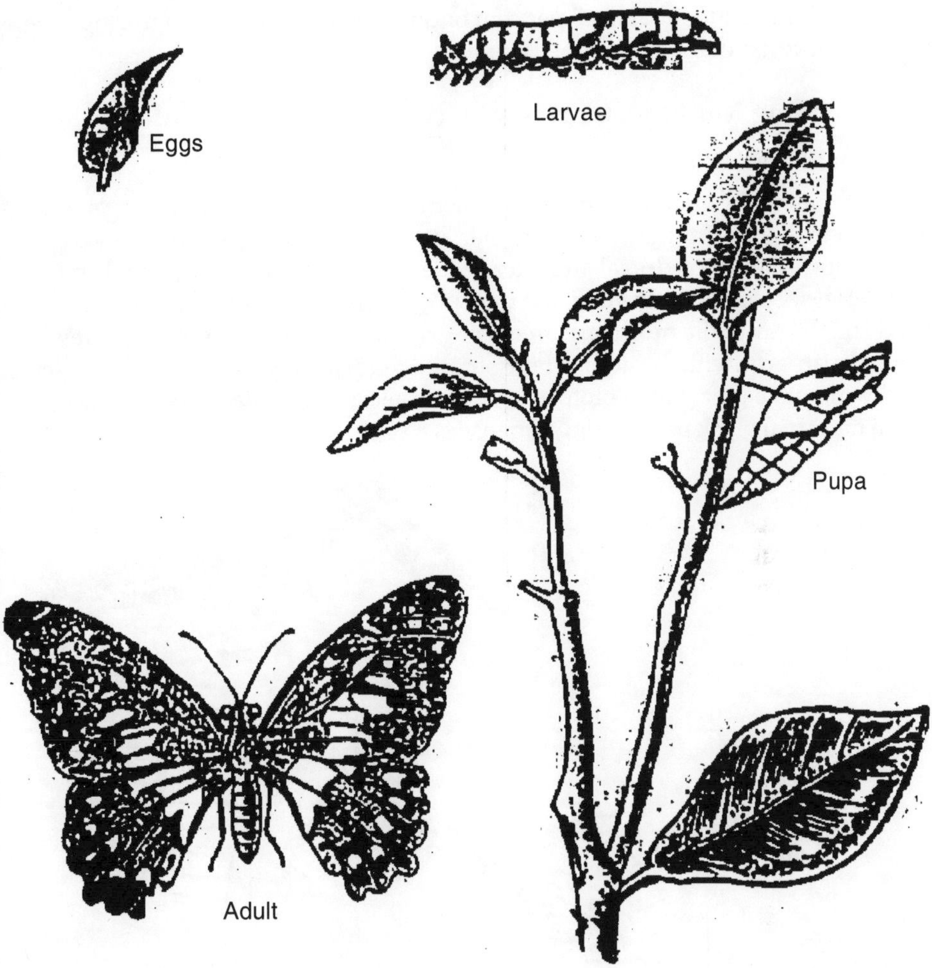

Figure 13.11: Lemon Butterfly

under surface. A female lays on an average 75-120 eggs. The eggs are placed singly or in groups of 2-5 eggs. Eggs hatch in 3-4 days during summer and 5-8 days in winter. The larval period lasts 8-16 days in the summer and about 4 weeks in winter. The mature larva spins a supporting girdle around its body and pupates on a twig. The pupal stage lasts about 8 days in summer and 9-11 days in winter. A male lives for 3-4 days and female for about a week. They passes through three or four generations in a year.

Management

1. Hand picking of various stages of the pest and their destruction especially in nurseries and new orchards helps to suppress the population of the pest.
2. Spraying of entomogenous bacteria, *Bacillus thuringiensis* @ 2 kg per ha and neem seed extract 3 per cent also gives quite high mortality of caterpillar.

3. Spray endosulfan 0.07 per cent or quinalphos 0.05 per cent during severe infestation *i.e.*, September.

Fruit Sucking Moth, *Ophideres* spp. (Lepidoptera: Noctuidae)

Marks of Identification

The freshly hatched caterpillars are slender, thread like 3-4 mm long and pale greenish in colour. Full grown caterpillar are 50-60 mm long, stouty, velvety blue with yellow pattern on dorsal and lateral sides and having a hump at anal end. The moths are large and stoutly build and their prominent palpi are turned upwards. The head and thorax of the moths are greenish-gray and abdomen is orange. Forewings are greenish-grey with numerous faint striated reddish lines and three rufous spots, hind wing have apical area blotched with rufous, a round black spot in the center and a marginal black band. Wing expanse is 80-110 mm.

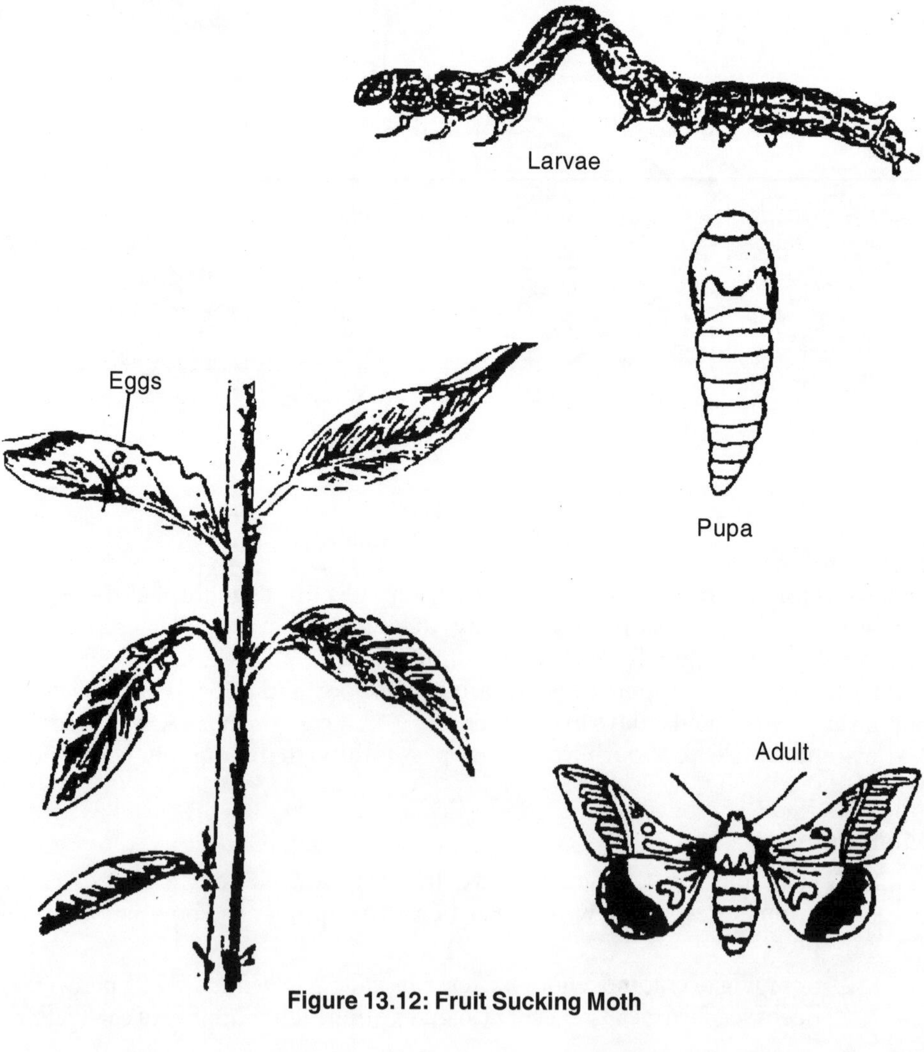

Figure 13.12: Fruit Sucking Moth

Nature of Damage

The moth punctures the ripening fruit for sucking juice with the help of its strong, piercing mouthparts. The damaged fruits soon start rotting as the punctures regions are easily infested with bacteria and fungi and ultimately the fruits drop prematurely. If the damaged squeezed, the juice spurts from the hole. In severe cases of infestation almost all the fruits are lost.

Life Cycle

The moths are nocturnal and are not seen during the day. They lay eggs on a number of wild plants and weeds, which are often found growing near citrus orchards. The eggs hatch in about 8-10 days within 24 hours of emergence, the young larvae start feeding on the foliage of host plants. A larva passes through five instars in 28-35 days. When full grown it makes a pupal case by webbing together pieces of leaves and soil particles. The pupa is thickset and is dark reddish brown. This stage lasts about 14-18 days. The moths on emergence fly to nearby orchards for feeding on fruit-juice.

Management

1. Destruction of alternate host plants in the vicinity of the orchard is advised to minimize the pest infestation.
2. Dispose the fallen fruits which attract the moths.
3. Creating smoke in the orchards after sunset may keep the pest away.
4. Install light trap to attract the moths and destroy the attracted moths.
5. Bagging the fruits to protect from the moth attack.
6. Kill moths with a bait containing gur 100 g+ vinegar 6.0 ml + malathion 10 ml+ water 1 litre. Wide-mouthed bottles (1 bottle per 10 trees) containing bait solution should be tied to the plants when the fruits are in unripe condition.
7. Spray tree with carbaryl 0.2 per cent at the time of maturity of fruits.

White Fly, *Dialeurodes citri* (Hemiptera: Aleurodidae)

Marks of Identification

The eyes are oval and pale yellow. The adult is a minute insect, measuring 1.02 to 1.52 mm. The males being smaller than the females. The antennae are six-segmented. The eyes are transparent, red and kidney shaped, with the lower half covered over with bristles. The head is somewhat pointed. The wings are more than twice in length of the body and extend beyond the tip of the abdomen. Both the wings and the body are completely covered with white waxy powder. The nymphs are pale yellow, with purple eyes and its body is marginally fringed with bristles.

Nature of Damage

The nymphs and adults reduce the plant vigour by sucking large quantities of cell sap. Severely infested leaves turn pale green and gradually become pale brown, get badly curled and are even shed. The nymphs also produce copious quantity of

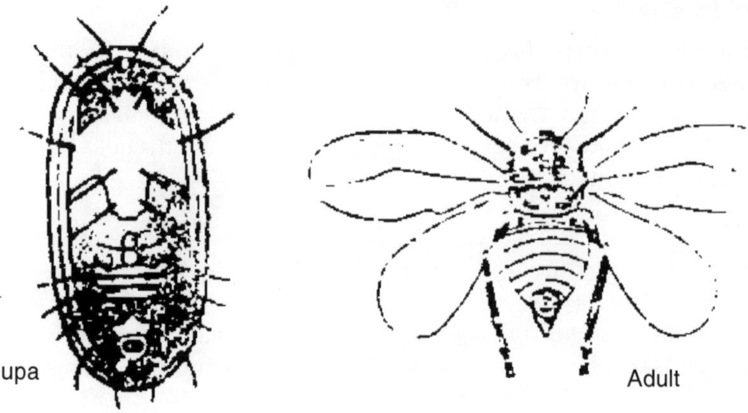

Figure 13.13: Mango White Fly

honeydew on which sooty mould develops covering the foliage with superficial black coating and interfering with photosynthesis of the plant. As a result, the growth of the tree is stunted and the affected trees produce few blossoms, most of which are shed and even the fruits that are formed are insipid.

Life Cycle

The female may lay 200 or more eggs singly on the underside of soft young leaves during its life of 7-10 days. It may take 10-20 days for the eggs to hatch. The young nymphs on emergence crawls about for a few hours and then settles on the succulent portion of the twig. The nymph is full-fed in 25-71 days and then change to pupa. The pupal period is long in summer and cold season where it lasts 114-159 days.

Management

1. For effective management of flies, close planting, water logging or any other stress condition should be avoided.
2. In case of localized infestation, the affected shoots should be clipped off and destroyed.
3. Spray methyl-o-demeton 0.025 per cent or dimethoate 0.03 per cent or malathion 0.05 per cent or monocrotophos 0.04 per cent or phosalone 0.07 per cent or endosulfan 0.07 per cent at peak activity of this pest.

Citrus Black Fly, *Aleurocanthus woglumi* (Hemiptera: Aleurodidae)

Marks of Identification

The male and female adults are 0.8 and 1.2 mm long, respectively. When freshly emerged, head and thorax are bright red, eyes reddish brown while antennae and legs are whitish. Within 24 hours the adults become slaty-bluish. Wing has black patches or whitish background. The full grown nymphs are 1-2 mm long and dark brown to shiny black in colour and conspicuously spiny, bordered by a white fringe of wax.

Nature of Damage

Both adults and nymphs suck plant sap, reducing the vitality of trees. It results in the curling of leaves and also the premature fall of flower buds and developing fruits.

Life Cycle

A single female may lay more than 100 eggs in spiral pattern on ventral side of leaves in her lifetime. There may be 16-22 eggs in a cluster. The eggs hatch in 7-14 days. The nymphs on emergence start feeding on cell sap. They pass through four nymphal instars and the nymphal stage is completed in 38-60 days. They pupate in leaf surface and the stage lasts 100-131 days. There are two distinct broods in a year.

Management

1. For effective management of flies, close planting, water logging or any other stress condition should be avoided.
2. In case of localized infestation, the affected shoots should be clipped off and destroyed.
3. First and second nymphal instars of the pest are more vulnerable to control hence the spray schedule should be undertaken at 50 per cent egg hatching stage that ensures maximum coverage of younger stage of the pest in two sprays to be effected at 15 days intervals with any of the insecticides namely monocrotophos 0.04 per cent, acephate 0.04 per cent, phosalone 0.05 per cent, dimethoate 0.06 per cent till drenching that ensure through coverage of the underside of leaves.

Bark Eating Caterpillar, *Indarbela quadrinotata* (Lepidoptera: Metarbelidae)

Details given in the pests of guava.

Mealy Bug, *Planococcus citri* (Hemiptera: Pseudococcidae)

Marks of Identification

The adult is a minute insect, eyes are transparent, red and kidney shaped with the lower half covered with bristle. The wings are more than twice the length of the body. Both the wings are covered with white waxy powder. The nymph is flat, oval, pale yellow in colour with purple eyes. The body of the nymph is marginally fringed with bristles and scale like in appearance.

Nature of Damage

The pest cause heavy damage to nursery and grown up plants. Damage is severe in spring and autumn. The leaves and tender shoots get deformed and twisted into knot and loops. The leaves become curled up. It feeds at stem ends of fruits and often causes heavy fruit drop. Clusters of white formations of bugs are found at the joints of twigs. It excretes plant sap and reduces tree vigour. It also excrete honeydew which invites sooty mould.

Life Cycle

This pest breeds continuously. Yellow eggs are produced in a loose colony of waxy filaments. About 600-800 eggs covered in an ovisac are deposited in a fortnight which hatches within 7 to 15 days. There are 3 moults of female nymphs and 4 moults of winged male nymphs completing 3 overlapping generations in a year.

Management

1. Chemical control of the pest includes spraying of dimethoate 150 ml + kerosene oil 250 ml in 100 litres of water or carbaryl 10 ml + oil 10 ml or malathion 20 ml in 10 litres of water.
2. The possibility of the use of nymphal and adult parasitoid, *Leptomastrix dactylopii* and the predator coccinellid *Cryptolaemous montrouzieri* and the Chrysopid, *Mallada boninensis* needs to be explored for the control of the pest.

Citrus Aphid, *Taxoptera citricidus* and *T. aurantii* (Hemiptera: Aphididae)

Nature of Damage

Adults and nymphs suck the sap from tender leaves and shoots devitalizing the plants. The pest secretes copious sugary solution on which sooty mould grows. Brown citrus aphid *Toxoptera citricidus*, is responsible for vectoring citrus disease 'Tristeza'.

Life Cycle

Citrus aphid produces about 5 young ones daily for a period of 1-3 weeks parthenogenetically which attain maturity in about 6 days. Daytime warm temperature during February- March favours the pest population buildup. Winged colonizing adults called stem mothers give birth directly to nymphs with no egg stage and no mating requires. Nymph of black aphid gives out red hoemolymph and brown aphid, the yellow on squasing.

Management

1. An Ichneumouid parasitoid, *Lipolexis scutellaris* and the predators like chrysopids, coccinellids, syrphids ate, feed on the pest.
2. Spray quinalphos or dimethoate or malathion 0.05 per cent at weekly intervals to control the pest effectively.

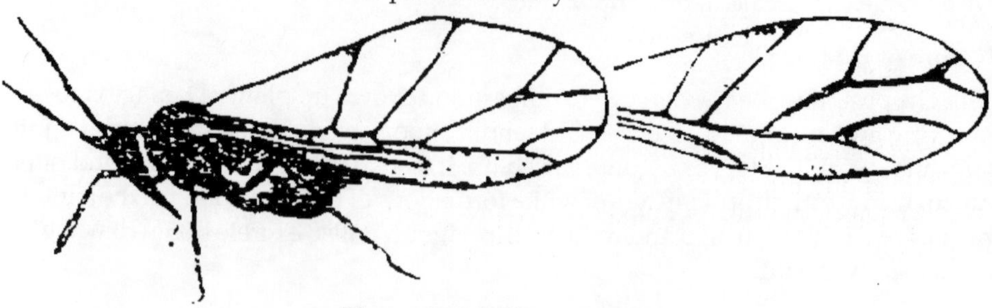

Figure 13.14: Citrus Aphid

Citrus Thrips, *Scirtothrips* spp. and *Heliothrips haemorrhaeodalis* (Thysanoptera: Thripidae)

Marks of Identification

The adults are slender yellowish brown in colour having apically pointed wings, and they measure about 1 mm in length. The females possess long narrow wings with the fore margin fringed with long hairs. The nymphs resemble the adults in shape and colour but are wingless and smaller in size.

Nature of Damage

The nymphs and adult rasp and suck the sap from fully developed flower and leaf buds, young and grown-up fruits and also the leaves. The leaves become cup shape and leathery. Two white lines parallel to leaf midrib and a whitish silvery ring around the fruit neck are characteristics of thrips infestation.

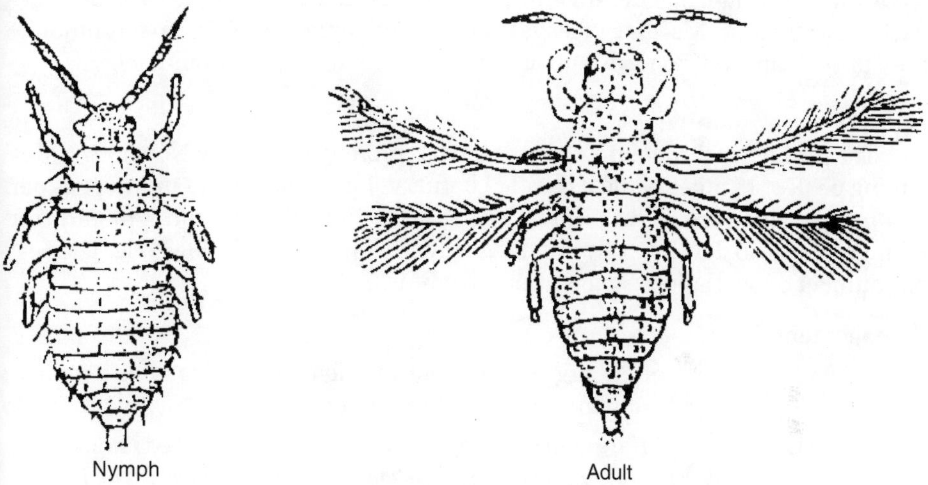

Figure 13.15: Citrus Thrips

Life Cycle

Males are practically unknown and the reproduction is by parthenogenesis. The female inserts eggs in the incisions made in leaf tissues which hatch in 2-7 days. The nymphal duration varies between 30 to 60 days depending upon the climatic conditions. It undergoes 3-4 instars to reach maturity. There are several generations in a year.

Management

1. Dimethoate or monocrotophos @ 1 ml per litre of water should be sprayed at bud burst stage and on berries and the surrounding vegetation should also be sprayed as the pest thrives on it.

Citrus Mite, *Eutetranychus orientalis* (Acarina: Tetranychidae)

Marks of Identification

The adult is small, plump and orange with thick deep brown patches on the dorsal side of its body and measures 0.33 mm in length. Its body is covered with prominent bristles, each born on a whitish tubercle. The antennae of the females are three segmented and are bright carmine pink. The newly hatched larva is light yellowish brown and has only three pairs of legs. The protonymph is orange brown and the duetonymph is orange brown with greenish ting.

Nature of Damage

The mite extracts cell sap from leaves and fruits. Mite feeding causes pale stripping on the upper surface of leaves which are not seen on the lower surface. In severe infestation the stripping enlarges to dry necrotic areas. But the stripping or silvering of green fruits disappears when the fruit changes colour. When large population feeds on maturing fruits the silvering may persist. Generally leaves drop and die-back of twigs starts. Mottling of leaves and chlorotic appearance due to multitude of grey unpleasant spots on fruits affecting the quality which is serious concern.

Life Cycle

The pest active throughout the year but has prominence during November- May having peak of 15 mites per leaf during February- June. Female lays upto 8 eggs per day mainly along the midrib of the leaf. It takes 10 to 15 days and 3 months for younger stages to develop fully and 17-20 days to complete entire life cycle. The pest has number of overlapping generations in a year.

Management

1. Water stress often aggravates mite problem. Trees are well irrigated particularly during the stress in late summer.
2. The most important natural enemies of citrus mite is a predacious mite, *Euseius hibisci* and the predators, *Agistemus* sp., and *Ambylesius hibisci.*
3. Foliar spray of dicofol @ 1.5 ml, or monocrotophos 1 ml or wettable sulphur 3 g per litre of water should be given as and when incidence is noticed.

Guava

Bark Eating Caterpillar, *Indarbela quadrinotata* (Lepidoptera: Metarbelidae)

Marks of Identification

Freshly hatched larvae are dirty brown while the full grown larvae have pale brown bodies with dark brown heads. The adults are pale brown moths with rufous head and thorax. The forewings are pale rufous with numerous dark rufous bands. The hindwings are fuscous.

Nature of Damage

On hatching the larvae bore into the bark or main stem. The larvae remain within the bored holes during day and come out at night to feed on the bark. They feed

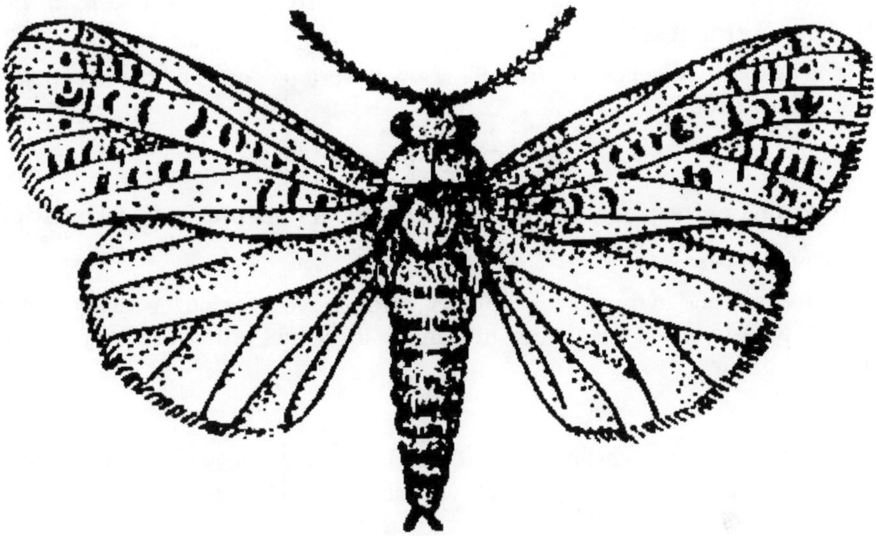

Figure 13.16: Bark Eating Caterpillar

sheltered under the silken galleries. The attacked trees show the presence of winding silken galleries full of frass and facial matter on the bark surface extending from the bored holes downwards. Generally one larva is seen in each hole but in case of severe infestation 15-30 larva may be bore the same tree. A severe infestation may result death of the attacked stem but not of the main trunk. There may be interference with the translocation of cell sap and thus, arresting of tree growth is noticed with the resultant reduction in its fruiting capacity.

Life Cycle

The moth starts laying eggs in clusters of 15-25 eggs each, under the loose bark of tree. A single female lays about 300-400 eggs. The egg hatch in 8-10 days. Larvae take as many as 9-11 months to complete development. Full grown larvae make a hole into the wood and pupate inside. The pupal stage is 3 to 4 weeks. Moths are short lived. There is only one generation in a year.

Management

1. Keep the orchard clean to prevent the infestation of this borer.
2. Kill the caterpillars mechanically by inserting an iron spike into the holes made by these caterpillars.
3. In case of severe infestation, clean the affected portion of the trunk or main stem and insert into the hole swab of cotton wool soaked in a 0.05 per cent emulsion of monocrotophos or dichlorvos (DDVP) and seal the hole with mud paste for killing the larvae in tunnels.

Fruit Fly, *Bactrocera dorsalis, B. diversus, B. nigrotibialis* and *B. zonatus* (Diptera: Tephritidae)

The former two species of fruit fly are very serious pest. The details are given under mango fruit fly.

Grapevine

Grapevine Leafhopper, *Erythroneura* spp. (Hemiptera: Cicadellidae)

Nature of Damage

Both adults and nymphs suck the cell sap from leaves causing the foliage blotched with tiny white spots. Under heavy infestation the leaves turn yellow or brown and fall from the vines.

Life Cycle

During the spring season the adults become active. Eggs are laid in leaf tissues and they hatch in about14 days. The nymphs are wingless and pale in colour and feed on the lower surface of the leaf. They moult five times. They become full develop in 3-5 weeks. There are 2-3 generations in year.

Management

1. When the infestation level increased spray fenitrothion 625 ml after rainy season.

Grapevine Thrips, *Rhipiphorothrips cruentatus* (Thysanoptera: Heliothripidae)

Marks of Identification

The adults are minute, being 1.4 mm long, blakish brown, with yellowish wings. The nymphs are yellowish brown in colour.

Nature of Damage

They feed in large number on plant leaves. The attacked leaves take a whitish hue, acquired a withered appearance, and then turn brown. The affected leaves ultimately curled up and drop off the plant. Such vines either do not bear fruit or the fruit drops off prematurely. Sometimes the fruits are also attacked resulting in scab formation on the berries. The peak infestation is during hot weather-May to July in South India, March to April and August to October in North India.

Life Cycle

Both sexual and parthenogenetic reproduction occur side by side, the progeny of the latter being always males. 2-6 eggs are laid per day. Incubation, pre-imaginal, pupal and total life cycle durations occupy 3-8, 11-22, 2-5 and 14-33 days, respectively. There are 5-11 generations in a year. The pest overwintering in pupal stage in soil.

Management

1. Remove grasses from orchard and prune infested leaves.

2. Spray malathion 500 ml in 500 litres of water per 100 vines, once before flowering and again after the fruit set.

Grapevine Girdler, *Sthenias grisator* (Coleoptera: Cerambycidae)

Marks of Identification

Adult beetles are about 24 mm long, grayish-brown with white and brown irregular markings resembling the bark colour, elytra have an elliptical grayish median spot and an eye-shaped patch.

Nature of Damage

The adult beetles emerge usually around July to August and maximum damage is caused during August to October. The beetles with their strong mandibles, ring or girdle the young green branches of the host plant. The girdle's branches dry up.

Life Cycle

The female deposit eggs in clusters of 2-4 underneath the bark of girdled branches. Incubation period is 8-10 days. The grub stage lasts for 7-8 months. The life cycle is completed in more than a year. There is only one generation in a year.

Management

1. Cut the affected branches below girdling point and burn the same.

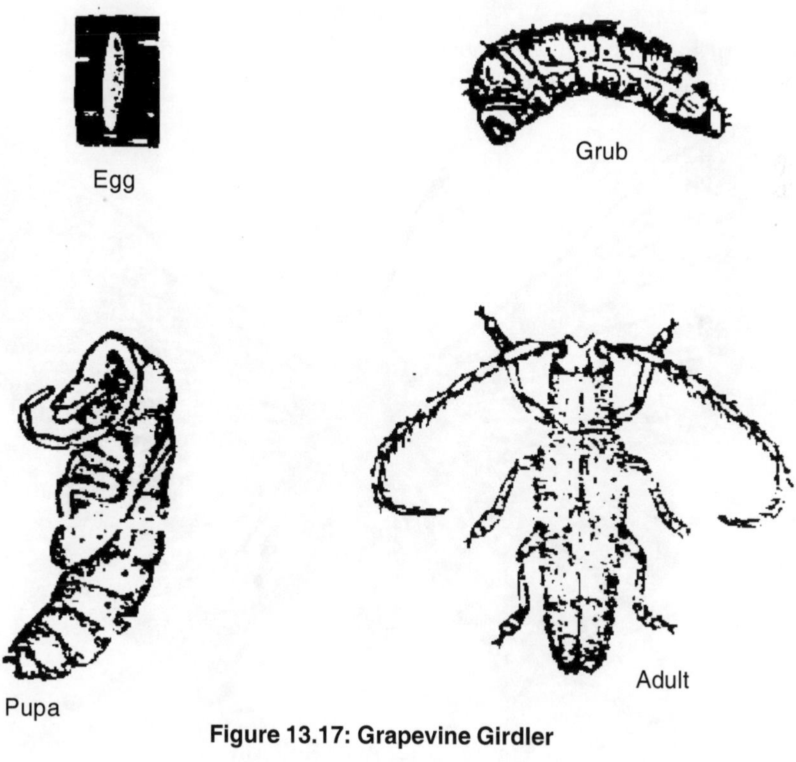

Figure 13.17: Grapevine Girdler

2. Hand collection and destruction of beetles also help in mitigating this pest.
3. Spray the vines with 1.5 litres of monocrotophos in 1250 litres of water.

Pomegranate

Anar Butterfly, *Virachola isocrates* (Lepidoptera: Lycaenidae)

Marks of Identification

Full grown caterpillars are stout, 17 to 20 mm long, dark brown in colour and have short hairs and whitish patches all over the body. Adult butterflies are medium sized, glossy-bluish-violet (male) to brownish-violet (females) with an orange patch on each wing. The wing expanse is 40-50 mm.

Nature of Damage

The caterpillars damage the fruit by feeding inside and ridding through the ripening seeds of pomegranate. As many as eight caterpillars may be found in a single fruit. The affected fruits ultimately fall off and give an offensive smell. This pest may cause 40-90 per cent damage to fruits.

Life Cycle

The pest breeds throughout the year on one fruit or the other. The female butterfly lays shiny white, oval shaped eggs singly on the calyx of flowers and on small fruits and leaves. The eggs hatch in 7-10 days and the young larvae bore into the developing fruits. They feed there for 18-47 days till they are full grown. Then they pupate inside the fruit, but sometime may pupate outside. The pupal period lasts 7-34 days. There are four overlapping generations in a year.

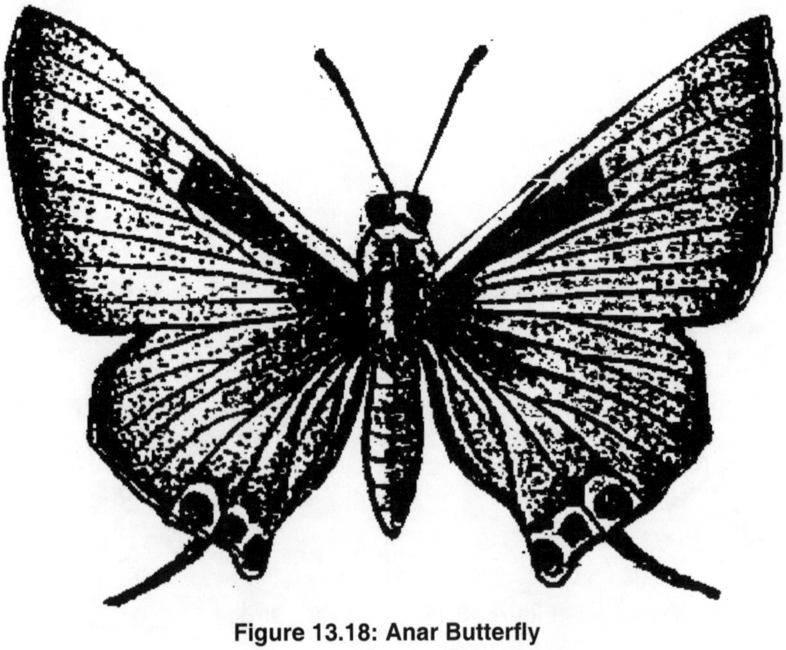

Figure 13.18: Anar Butterfly

Management

1. Collect and destroy the fallen infested fruits to prevent build up of the pest.
2. If the number of fruit tree is limited, bagging of the fruit before maturity is advised to reduce the fruit damage.
3. Spray endosulfan 0.07 per cent or monocrotophos 0.04 per cent 2 to 3 times at 15 days interval commencing from infestation of fruit borer.

Ber

Fruit Fly, *Carpomyia vasuviana* (Diptera: Tephritidae)

Marks of Identification

The adults smaller than the housefly, are brownish yellow, with brown longitudinal strips on the thorax, being surrounded on the sides and the back with black spots. Wings are hyaline, transparent with grayish brown spots and there are bristly hairs on the tip of the abdomen. Full grown maggots are amphineustic, cretaceous or creamy white and about 6 mm long.

Nature of Damage

The female flies puncture the ripening fruits and lay their eggs inside the epidermis. The development around the punctures is arrested and this causes the deformation of fruit. On hatching maggots feed on fleshy and juicy pulp. The infested fruit turn dark brown, rotten near the stones and emit a strong smell. One maggot is sufficient to destroy the entire fruit. Fleshy varieties of ber more seriously damages than the less fleshy once. Late maturing fruits are destroyed almost entirely.

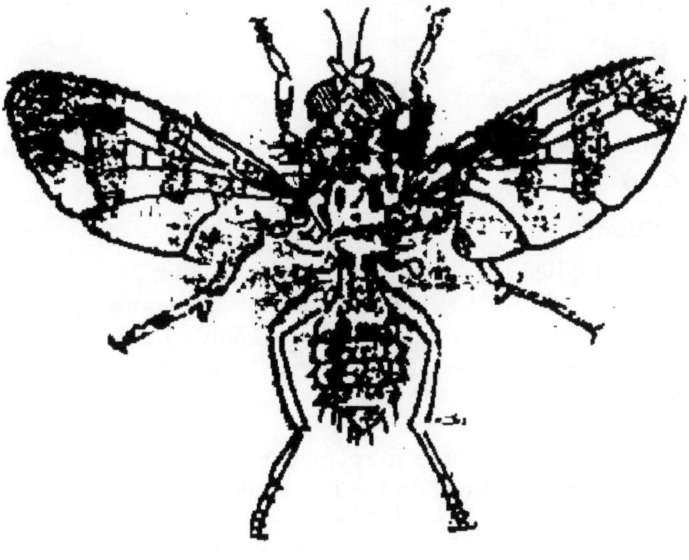

Figure 13.19: Ber Fruit Fly

Life Cycle

The female fly make cavity in the skin of fruit and lay 12 to 18 eggs singly or in groups of 2 to 4. The eggs hatch in 2-3 days and the maggots feed on the flesh of the fruit, making galleries towards the center. The larvae are full grown in 7-10 days and they come out of the fruit by cutting one or two holes in the skin. They move away, making jumps of 15-26 cm and reach a suitable place to pupate generally 6-15 cm below the soil surface. The pupal stage lasts 14-30 days and the shortest life cycle, from egg to the adult emergence is completed in 24 days. There are 2-3 broods in a year.

Management

1. Collect and destroy the fallen infested fruits at alternate days.
2. Rake the soil around the trees during summer to expose the pupae to sun heat and natural enemies.
3. To escape egg laying on fruit, do not allow the fruits to ripe on the tree, harvest at green and firm stage.
4. Grow resistant/tolerant varieties *viz.*, Safada, Sanaur-1, Rohtaki Gola, Katha, Desi Alwar, Tikadi, etc.
5. Install methyl eugenol traps (0.05 ml or 4 drops each of methyl eugenol and dichlorvos to be recharged at weekly interval) in ber orchard to trap and kill the male fruit flies.
6. Three spray of monocrotophos 0.03 per cent control the infestation effectively. The first spray applied when ber fruit attain peanut size.
7. The insecticides, malathion 0.07 per cent, neemark 0.1 per cent, dichlorovos 0.03 per cent and quinalphos 0.05 per cent are also effective against fruit fly.

Banana

Banana Corm Weevil, *Cosmopolites sordidus* (Coleoptera: Curculionidae)

Marks of Identification

Adult weevils are 10-13 mm long shiny black in colour, having fairly long and curved snout and short elytra striated longitudinally. Though these wings are functional the weevils seldom fly. The grubs are creamy white, stout, fleshy, highly wrinkled and legless with spiny-shaped, 8-12 mm long body.

Nature of Damage

Infestation at early stage reduces the plant vigour. Early stage damage includes sick appearance and yellow lines on the leaf whereas in the advanced stage of infestation plant show tapering of stem at the crown region, reduction in leaf size, poor bunch formation and choked throat appearance of bunch due to riddling by the grubs inside the corm.

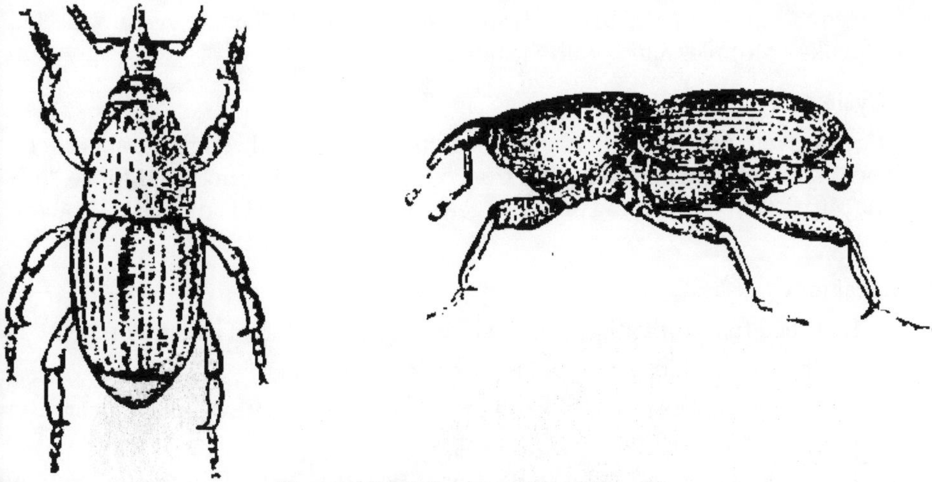

Figure 13.20: Banana Corm Weevil

Life Cycle

The adult weevil lay eggs singly into the corm. Eggs are laid superficially in the surface and crevices of the corm. Incubation period is 7-14 days. Larval stage lasts for two to six weeks. Pupation is completed in a week. Adult live upto two years and can remain without food for 6 months.

Management

1. Weed free cultivation is essential to avoid the spread of weevils.
2. Remove the old and dried leaves.
3. Remove the pseudostem after harvest and treat it with carbaryl (2 g per litre).
4. Paring to remove the adult weevil and grubs from the corm. Apply carbofuran 3 G (40 g per plant) before planting.
5. Before planting, treat the suckers with monocrotophos (14 ml in one litre of water) for 30 minutes.
6. Collect adult weevils by using banana pseudostem traps both- longitudinal split trap (30 cm length) and disc-on-stump trap (100 per ha).

Banana Stem Weevil, *Odoiporus longicollis* (Coleoptera: Curculionidae)

Marks of Identifications

Adult weevils are chocolate to black in colour, 23-28 mm long with pointed head. Grubs are apodous, soft, fleshy, wrinkled and sparsely covered with brown hairs of varying lengths. Pupae are exarate and pale yellow in colour.

Nature of Damage

Presence of small holes and jelly exudation on the stem indicates the grub activity inside the stem. The apodous larvae are responsible for riddling of the pseudostem and causing serious damage. In the advanced stage of infestation the severely affected plant breaks or topples along with the bunch.

Life Cycle

Eggs are laid inside the air chambers of the leaf sheath through the slits cut on leaf sheath. Incubation period ranges from 5-8 days. Larval period lasts for 26 days and the pupal period including pre-emergence resting period of adult lasts for about 20-24 days.

Management

1. Weed free cultivation helps in reducing the spread of infestation.
2. Remove the old and dried pseudostems of banana.
3. After harvest remove the pseudostem and treat it with insecticide, carbaryl (2 g per litre) to kill the egg laying weevils.
4. Collect adult weevils by using banana pseudonym traps both- longitudinal split trap (30 cm length) and disc-on-stump trap (40 per ha).

Jack Fruit

White Tailed Mealy Bug, *Ferrisia virgata* (Hemiptera: Pseudococcidae)

Marks of Identification

The adult female are dark castaneous and 4.0-4.5 mm long. They are covered with a sticky cretaceous white ovisac. The nymphs are deep chocolate in colour, having their dorsum covered thinly with a whitish mealy material.

Nature of Damage

This pest is very active during dry season. It feeds on leaves and tender shoots and during dry weather it moves down and inhabit the roots. A prolonged period of drought may result in a severe outbreak of this pest. The peak period of its activity is August to November, when the affected parts turn yellow, wither and ultimately die away.

Life Cycle

The eggs are enclosed in ovisac. The female lays 200-400 eggs which are cylindrically rounded, flat at both ends and chestnut in colour. The eggs hatch in 7-10 days. Male and female nymphs take 15 and 20 days to complete development, respectively.

Management

1. Spray 0.05 per cent monocrotophos or diazinon.

Leaf Webber, *Perina nuda* (Lepidoptera: Lymantriidae)

Marks of Identification

Full grown caterpillars are 22-25 mm long, having short erect tufts of dusky grey to brownish hairs. In the moths there is extreme sexual dimorphism. Male moths are smaller than the female moths. The male moths have half ochreous and half transparent forewings with a wing expanse of 33-36 mm, and the female is dull ochreous white in colour having 38-42 mm wing expanse.

Nature of Damage

The caterpillar folds or webs the leaves together and feed on the green matter causing extensive defoliation.

Life Cycle

A single female lays 60 to 400 eggs. They hatch in 4-6 days. Larval and pupal periods are 16-20 and 5-9 days, respectively. The adult longevity is 3–11 days and the total life-cycle is completed in 27-39 days.

Management

1. Hand picking and mechanical destruction of caterpillars in the initial stage of attack prevent further build up of population of leaf webber.
2. In case of severe infestation spray 750 ml monocrotophos or 1.50 litre chlorpyriphos 20 EC in 625 litre of water per ha. Spray should be directed on the upper surface of foliage.

Pineapple

Pineapple Thrips, *Thrips tabaci* (Thysanoptera: Thripidae)

Marks of Identification

Adults are small, about one mm long, yellowish-grey to brown in colour with darker transverse bands across the thorax and abdomen. Forewings are shaped and possess short setae.

Nature of Damage

Nymphs and adults rasp the epidermis of leaves and imbibe the oozing sap, thereby devitalizing and distorting the leaves and affecting the growth adversely. In case of severe infestation the leaves show silvery sheen and bear small spots of faecal matter. In many parts of the world this thrips has also been recorded as a vector transmitting yellow spot virus disease of both plant and fruits.

Life Cycle

Reproduction appears to be thelytokous parthenogenetic. Males are extremely rare, the ratio being 3000 females to one male. Eggs are inserted into tissues of green leaf through a cut made by the ovipositor. A single female lays 40-50 eggs. These hatches in 4-9 days and the entire life cycle is completed in 10-13 days in summer and 21 days in winter season.

Management

1. Removal of alternate host plants near pineapple orchards.
2. Spraying 0.04 per cent diazinon.

Slug Caterpillar, *Latoia lepida* (Lepidoptera: Limacodidae)

Marks of Identification

The full grown larvae is flat, fleshy and greenish in colour with white lines on the body. It is curved with spines having red or blue tips. The adult moth is short and stout with green and brown forewings.

Nature of Damage

The young caterpillars feed gregariously by scrapping the under surface of the leaves. The continuous feeding leads to loss of sap from plant tissues ultimately reduce the vitality of plants and the affected leaves dry up.

Life Cycle

The female moths lay eggs in batches of 15-35 on the under surface of the leaves. The eggs hatch in 7 days. On hatching the young larvae start feeding in clusters on the under surface of leaves. The larva passes through five instars and is full-developed in 40-47 days. It pupates in silken cocoon in stems. The pupal stage lasts for 20-28 days and the life cycle is completed in about 10 weeks.

Management

1. Hand picking and mechanical destruction of caterpillars at early stage of infestation.
2. Dusting with methyl parathion (5 per cent) is effectively and should be done when the caterpillar are still feeding gregariously.

Date Palm

Rhinoceros beetle, *Oryctes rhinoceros* (Coleoptera: Curculionidae)

Marks of Identification

The adult beetles are robust, 35-50 mm long, glossy-dark-brown to black dorsally and reddish brown ventrally. The pygidium in males is rather long, smooth and rounded while in case of females, it is short hairy and relatively pointed. Grubs are 80-100 mm long, soft, fleshy and wrinkled. Pupae are uniformly yellowish-brown.

Nature of Damage

Beetles are the destructive stage of this pest. They are nocturnal in habit. At night these beetles feed on the crowns of palms by boring into the unopened tender fronds and tunnel downwards feeding on soft tissues of the growing point. As a result, the growth of the tree is arrested. Young trees are more prone to attack by this pest and these trees, when attacked ultimately wither and die away. The pest causes maximum damage during monsoon.

Life Cycle

Eggs are laid in rotting vegetation, manure or compost heaps. These hatch in 8-14 days. Grub period is 47-191 days and pupation takes place in the soil, 150-600 mm deep. Pupal period is 14-29 days and the adult remain in cocoon for 5-26 days till

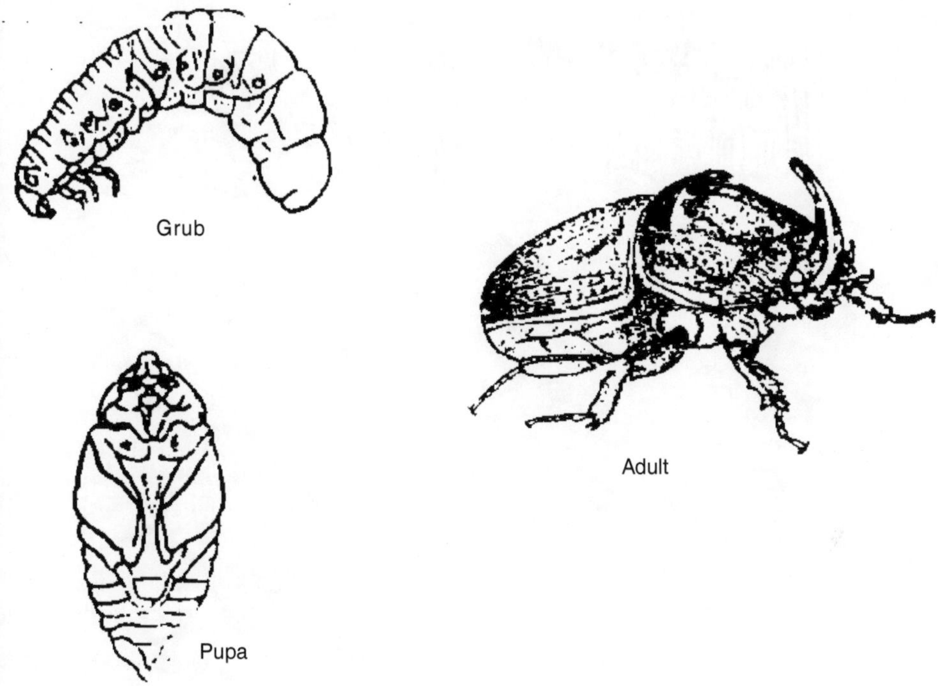

Figure 13.21: Rhinoceros Beetle

they become sexually mature. The life cycle is completed in 100-260 days and the adult live for 76-219 days.

Management

1. Keep the orchards clean and adopt all sort of sanitary practices.
2. All potential breeding sites *viz.*, compost heaps, refuse dumps, dung hills etc., must be destroyed constantly in and around the orchards by deep burying or burning. These breeding sites may be dusted with methyl parathion (5 per cent).

Red Palm Weevil, *Rhynchophorus ferrugineus* (Coleoptera: Curculionidae)

Marks of Identification

Adult weevils are reddish brown in colour with red spots on thorax, cylindrical, 32-36 mm long, flattened with long and slightly curved snout. The grubs are fleshy, wrinkled, transversely apodous, pale yellow when freshly formed, later becoming light brown and full grown grubs are 50-60 mm long.

Nature of Damage

The grubs are voracious feeders, once enter inside the trunk they seldom come out. They burrow in hard woody portion and tunnel through the tissues in all

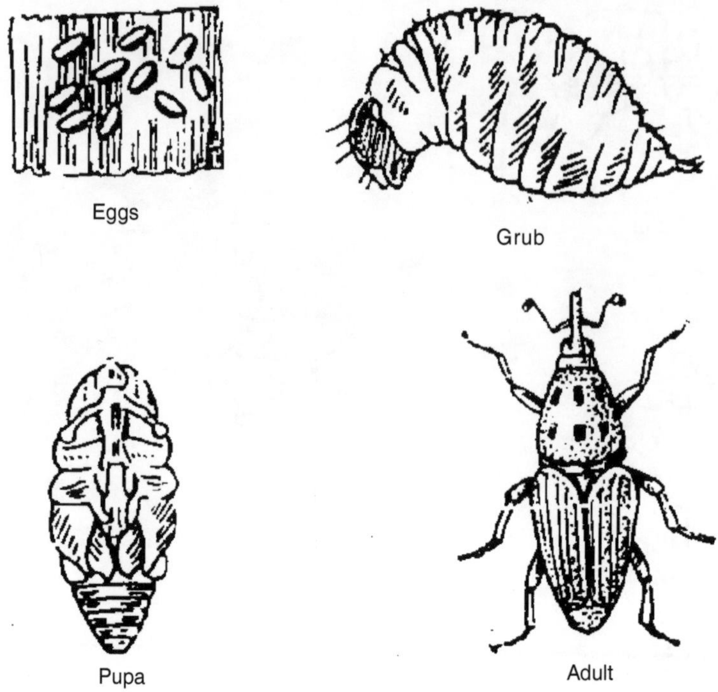

Figure 13.22: Red Palm Weevil

directions and feed on soft succulent tissues, discarding all fibrous material. In case of severe infestation, inside of trunk is completely hollowed and filled with decaying rubbish. As a result of this infestation, in case of young palms, the tops wither, rot and give out offensive smell, while in older trees, the top portion of trunk bend and ultimately breaks at the bend. A thick reddish brown viscous fluid as also chewed up and discarded fiber pieces are thrown out from the holes.

Life Cycle

The pest breeds throughout the year and all the stages are passed on the host tree. A female lays 76-355 eggs during its life span of 2-3 months. Eggs are laid singly in soft tissues at the base of leaf sheaths or in the cuts or wounds on the tree trunk. These hatch in 2-5 days. Grub and pupal period are 24-61 and 18-34 days, respectively and the entire life cycle is completed in 50-90 days. After emergence the weevils remain in cocoons for 4-17 days, till they become sexually mature.

Management

1. Maintenance of orchard sanitation, prompt destruction of dead and badly infested plants.
2. Filling up the leaf axils with 5 per cent methyl parathion + saw dust or sand (1:1 ratio) three times in a year. This will kill the female weevils that usually rest in leaf axils.

3. Prevent the occurrence of cuts, scars or wounds on tree trunks. As and when they appear, paint the same immediately with lime or coal tar. This will prevent the weevils from laying eggs on these trees.
4. If the attack of the pest is detected in the initial stage, remove and kill the grubs, mechanically.

Date Palm Scale, *Aspidiotus destructor* (Hemiptera: Diaspididae)

Marks of Identification

The scales of male are on an average 1.2 mm long and white in colour and the nymphs below are pale yellow and invisible. The female scales are brown white margin, more or less oval, flat and about 1.5 mm long. The mature females are creamy yellow and when gravid, these gradually become darker and darker till wine-red. Crawlers are also rose coloured and latter become light violet.

Nature of Damage

The pest prefers shade and invariably infests shaded pinnules and the main veins of the fronds. In severe cases even the fruits are attacked. Incidence is higher in trees where fruits have been wrapped with cloth as protection against solar radiation. Young trees grown in shade or those that are irrigated regularly suffer comparatively more from ravages of this pest. The continuous yellow incrustations of this scale on the leaflets, interfere with photosynthesis, respiration and transpiration activities of the trees which ultimately affect adversely the normal growth of the trees and development of fruits, the fruit become dry deformed and do not mature.

Life Cycle

An oviparous female may lay 6-10 eggs in her life time. The eggs remain under the female scales and hatch in 8-18 days. The nymphs become mature females normally in 30-40 days. Male nymphs have to pass through pre pupal and pupal stages before becoming winged or micropterous males. The life cycle varies from 50-60 days in summer to 75-85 days in spring.

Management

1. Good crop management restricts the pest development.
2. Remove and burn all the infested leaflets in initial stage of attack. This is possible only when the trees are young. In case of older trees, remove all the fronds except those at the very top and pour hot salty water over the crown and remaining leaves.

Jamun

Bark Eating Caterpillar, *Indarbela quadrinotata* (Lepidoptera: Metarblidae)

Marks of Identification

Freshly hatched larvae are dirty brown while the full grown larvae have pale brown bodies with dark brown head. The adults are pale brown moths with rufous head and thorax. The forewings are pale rufous with numerous dark rufous bands. The hindwings are fuscous.

Nature of Damage

On hatching the larvae bore into the bark or main stem. The larvae remain within the bored holes during day and come out at night to feed on the bark. They feed sheltered under the silken galleries. The attacked trees show the presence of winding silken galleries full of frass and facial matter on the bark surface extending from the bored holes downwards. Generally one larva is seen in each hole but in case of severe infestation 15-30 larva may be bore the same tree. A severe infestation may result death of the attacked stem but not of the main trunk. There may be interference with the translocation of cell sap and thus, arresting of tree growth is noticed with the resultant reduction in its fruiting capacity.

Life Cycle

The moth starts laying eggs in clusters of 15-25 eggs each, under the loose bark of tree. A single female lays about 300-400 eggs. The egg hath in 8-10 days. Larvae take as many as 9-11 months to complete development. Full grown larvae make a hole into the wood and pupate inside. The pupal stage is 3 to 4 weeks. Moths are short lived. There is only one generation in a year.

Management

1. Keep the orchard clean to prevent the infestation of this borer.
2. Kill the caterpillars mechanically by inserting an iron spike into the holes made by these caterpillars.
3. In case of severe infestation, clean the affected portion of the trunk or main stem and insert into the hole swab of cotton wool soaked in a 0.05 per cent emulsion of monocrotophos or dichlorvos and seal the hole with mud paste for killing the larvae in tunnels.

Tamarind

Chafer Beetle, *Holotrichia insularis* (Coleoptera: Scarabaeidae)

Marks of Identification

Adults are brownish-black, convex beetles, 18-20 mm long and 8-10 mm wide. Full grown grubs are dirty white, fleshy, curved, 38-44 mm long and 6-9 mm wide.

Nature of Damage

Adult beetles feed on foliage and in case of severe attack the seedlings and young trees may be completely defoliated. If the attack is heavy the saplings are killed outright whereas the young plants wither and gradually dry up.

Life Cycle

Mating usually takes place after the females have fed for about a fortnight. A single female lays 40-65 eggs. Eggs are laid in moist sandy soil, 30-150 mm deep. On hatching, the grubs feed on roots. Pre-oviposition and incubation period last for 4-6 and 8-12 days, respectively. While grub and pupal stages occupy 11-16 and 2-3 weeks, respectively. There is only one generation in a year. Pupae and adults hibernate in the soil from November to June.

Management

1. Plough around the trees, during winter to expose and kill the hibernating pupae and adults.
2. Soil treatment with phorate 10G 50-100 g per tree.
3. To kill the adults, spray 0.1 per cent fenitrothion especially during breeding season.

Tamarind Fruit Borer, *Phycita orthoclina* (Lepidoptera: Phycitidae)

Marks of Identification

The moths are small, delicate insects, having elongate forewings. The hindwings are broad and bear hairs on the dorsal side. The full-grown larva is cylindrical, pink and measure 14mm long.

Nature of Damage

The larvae feed on the pulp and their causing, excrements and webbing, render the fruit unfit for human consumption.

Life Cycle

The pest is active during December–April. The female moth lays flat, oval shaped and white eggs, singly on the pulpy portion inside the round shelled pods, through cracks and crevices found on their surface. A female on an average lays 190 eggs. Egg hatch in 4-5 days and larvae enter into the fruit pulp and feed there by making silken web. Larval stage is completed in 27-40 days. The larvae makes a silken cocoon inside the infested pod and pupate there. Pupal stage lasts for 6-8 days.

Management

1. Spray monocrotophos 0.04 per cent to control the pest.

Papaya

Ak Grasshopper, *Poekilocerus pictus* (Orthoptera: Acrididae)

Marks of Identification

Nymphs are yellowish-white with orange and black strips and dots all over their bodies. Adults are stout, yellowish with broad bluish-green strips on head and thorax, antennae bluish black with yellow rings, abdomen yellowish with transverse bluish-black bands. Forewings are bluish-green with yellow veins and reticulations, hindwings hyaline.

Nature of Damage

Both adult and nymphs feed voraciously on leaves and skeletonize them. In case of severe infestation even the bark of the trees is not spared.

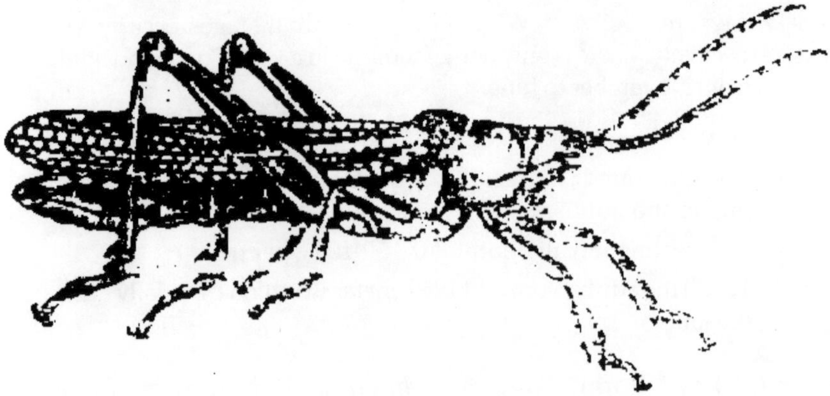

Figure 13.23: Ak Grasshopper

Life Cycle

Mating lasts for 5-7 hours and pre-oviposition period is 3-4 weeks. The female thrusts its abdomen into the soil and lays about 150-180 eggs at a depth of 120-200 mm depending upon the texture of the soil. There are atleast two broods in a year- a short one with incubation period one month (June- July) and nymphal period two months and a long generation when eggs laid during September to November over winter and hatch around the end of March or early April and become adults in another 2.5 months.

Management

1. Grasshoppers can be managed by spraying 0.1 per cent malathion or dusting malathion 5 per cent @ 50 kg per ha.

Sapota

Bud Borer, *Anarsia achrasella* (Lepidoptera: Gelechiidae)

Marks of Identification

The forewings of the adult moth are dark ashy grey in colour while hindwings are fringed with yellowish colour. A tuft of yellowish silky hairs, presents on the last abdominal segment in case of male and projections on labial palp in female. The full grown larva is pink in colour.

Nature of Damage

The larvae bore into flower buds and feed on ovary and petals, adversely affecting the production. Sometime they feed on leaves also. The pest is active throughout the year with peak in the beginning of monsoon.

Life Cycle

Female lay eggs on bud surface and stock of bud. Incubation period is 4-6 days. The larvae pass through four instars. Larval period is 11-13 days. The freshly formed pupa is brick red in colour. The pre-pupal and pupal period are 1-2 and 5-6 days, respectively. The female lays on an average 39-63 eggs with an average oviposition period of 2-4 days. The longevity of male and female moths are 4-5 and 4-6 days, respectively.

Management

1. Remove and destroy all the infested leaves, buds and fruits to reduce the infestation.
2. Spray monocrotophos 0.05 per cent or dichlorovos 0.03 per cent or endosulfan 0.07 per cent or fenthion 0.1 per cent as and when flush of flower bud set on the chiku trees.
3. Installation of traps containing black *tulsi* (*Ocimum sanctum*) leaves extract along with dichlorovos traps per two Sapota trees during April to September has been recommended which can attract the male moths of bud borer.
4. The black *tulsi* leaves extract poison bait trap installed at 4.5 cm height from ground level of peripheral canopy of chiku tree followed by three sprays of endosulfan (0.07 per cent) at 15 days interval starting from bud initiation is recommended for effective control of this pest.
5. Two spray of nimbecidine 0.3 per cent (neem based formulation) at 20 days intervals starting from March are also found effective for the control of chiku bud borer.

Chiku Moth, *Nephopteryx eugraphella* (Lepidoptera: Pyralidae)

Marks of Identification

It is a major pest of chiku and occurs widely in India. The moth is gray in colour, forewings slightly suffused with brown and irritated with black, Hindwings are semi-hyaline. The larva is 25 mm long, slender in body shape, and is pinkish with few longitudinal lines on the dorsal surface.

Nature of Damage

The larva causes considerable damage to young terminals affecting young buds and either stigma and style or corolla of the chiku bud and feed on ovary and inner parts of the bud, and then it move on the next buds thus, damaging many of them. Generally first larval instar completes within the flower bud and thereafter it comes out of the bud and produces a parchment. Bigger larva feeds voraciously on tender leaves, buds and young fruits causing considerable losses. The infestation is easily detectable by the presence of cluster of dried leaves hanging on webbed shoot.

Life Cycle

With the onset of spring season, the female moths start laying pale-yellow, oval shaped eggs singly or in batches of 2 or 3, on leaves and buds of young shoots. A female may lay as many as 374 eggs in 7 days. The eggs hatch in 2-11 days. The larvae feed for 13-60 days and complete development. They undergo pupation in the leaf webs and this stage is completed in 8-29 days. The life cycle is completed in 26-92 days depending upon the varying environmental conditions. There are 7-9 generations in a year.

Management

1. Remove and destroy all the infested leaves, buds and fruits to reduce the infestation.
2. Installation of traps containing black *tulsi* (*Ocimum sanctum*) leaves extract along with dichlorovos traps per two Sapota trees during April to September has been recommended which can attract the male moths of bud borer.
3. Spray monocrotophos 0.04 per cent or endosulfan 0.07 per cent following economic threshold level (one larva per twig) of this pest.

Sapota Fruit Fly, *Bactrocera* spp. (Diptera: Tephritidae)

Marks of Identification

The adult fly is stout and measures 14 mm across the wings and 5 mm in maximum length. The flies are strong fliers and can fly up to two kilometers in search of food. The fly is brown or dark brown in colour with hyaline wings and yellow legs. The thorax is ferruginous without yellow middle stripe. The abdomen is conical in shape and dark brown in colour. The young maggot is white and translucent.

Nature of Damage

The dark puncture caused by the oviposition of adult fly is not very conspicuous as its colour blends with the dark green colour of the fruit. It is very clearly visible in some yellow and pale brown varieties. The maggots on hatching feed on the pulp of the fruit for few days and a brown rotten parch appears on the fruit surface. The mesocarp becomes dirty brown. Infested fruits finally fall on the ground. The fruit is affected from late April to June.

Life Cycle

A female lays, on an average 50 eggs in one month. Eggs hatch in 3-10 days. As maggots develop, they pass through 3 stages in the ripening pulp and are full-grown in 6-29 days. They leave fruit and move away by jumping in little hope. On reaching a suitable place, they burry themselves into soil and pupate. In 6-44 days, they emerge as flies and reach ripe fruit for further multiplication. Life cycle is completed in 2-13 weeks and many generations *i.e.*, 10-12 are completed in a year. Flies are present in the field all through the year.

Management

1. To prevent the attack of fruit fly, harvesting of the fruits before ripening, and collection and destruction of infested fruits have been advised.
2. Install six methyl eugenol traps (0.056 ml or 4 drops each of methyl eugenol and dichlorovos to recharge at weekly interval) per hectare at a height of four feet placed at equal distance in Sapota orchard to trap and kill the male fruit flies.

Phalsa

Plum Hairy Caterpillar, *Euproctis fraterna* (Lepidoptera: Lymentriidae)

Marks of Identification

Adults are yellow moths with pale transverse lines on forewings, wing expanse is 24-28 mm in male and 30-38 mm in female. Full grown caterpillars are 35-40 mm long and have red head and darkish-brown body with which hairs on the head and a tuft of long hairs at anal end.

Nature of Damage

The caterpillar feed gregariously on leaf lamina skeletonizing the leaf completely. Later the caterpillars segregate the leaves. In case of severe infestation, the entire tree may be defoliated.

Life Cycle

Eggs are laid in clusters on ventral surface of leaves and are covered with yellow hairs. A female lays 150-300 eggs. Incubation, larval and pupal period last for 4-10, 13-29, and 9-25 days, respectively. There are three generations in a year.

Management

1. The freshly hatched larvae feed gregariously so, hand picking on small scale is quite effective.
2. Dust methyl parathion (5 per cent) for the control of this pest.

Custard Apple

Mealy Bugs, *Ferrisia virgata, Maconellicoccus hirsutus* (Hemiptera: Pseudococcidae)

Marks of Identification

The eggs are buff to light yellow in colour and oval to cylindrical in shape having round end, measured 0.34 mm in length and 0.17 mm in width. Female is apterous, soft bodied, elongate oval in shape and deep brown in colour with hair of prominent caudal filaments at posterior end and body is covered with a number of waxy filaments. Adult males are slender, delicate, elongate in shape and smoky brown in colour with a single pair of mesothoracic wings and two long waxy anal filaments. The freshly hatched adult female measured 1.96 mm in length and 0.90 mm in breadth while male measured 1.16 mm in length from head to tip of abdomen and 2.34 mm in breadth with wing expansion.

Nature of Damage

The infestation of this pest first appears on the seedlings grew up in the orchard. They infest leaves, shoots, buds and mainly fruits and suck the cell sap there from. As a result heavily affected development, fruits are reduced in size or get dropped. These insects also secrete honeydew on which a non-parasitic fungus, *Capnodium* species grows very rapidly, covering the fruits and plant parts with sooty mould. This black coating in turn interferes with the photosynthetic activities of the tree and as a result, its growth and fruiting capacity are adversely affected. The honeydew also attracts the ants which help in dissemination of the pest from tree to tree. Infested mature fruits also reduce their market value.

Life Cycle

They are active and mobile throughout the life. Reproduction is sexual as well as parthenogenetic, the later being more common. Mating takes place only once and a fertilized female lays 100-300 eggs in 3-4 weeks. The egg masses remain under the females till the young ones hatch out. Incubation period is about 3-4 hours. The development period of male and female nymphs varies from 31 to 57 and 26 to 47 days, respectively. Longevity of males is only 1-3 days while that of females extends from 36-53 days.

Management

1. To prevent the carry over of the pests, collect and destroy all the fallen infested fruits periodically and cut the branches portion being attacked between tree and also ground level.
2. Remove the seedlings and weeds, supporting the initial infestation, from the orchard.
3. Plough around the trees during summer to expose and kill the hibernated insect to natural enemies and sun heat.
4. Apply methyl parathion 2 per cent dust in the racked soil around the tree trunk to destroy the emerging crawlers.
5. Spray methyl parathion 0.05 per cent on stem to kill the climbing crawlers.
6. Remove and destroy the affected fruits and twigs to minimize the pest population.
7. Foliar application of the chlorpyriphos 0.05 per cent or quinalphos 0.05 per cent should be done as and when pest appears on tree.

Coconut

Leaf Eating Caterpillar, *Opisina arenosella* (Lepidoptera: Xyloryctidae)

Marks of Identification

The moth is ash grey in colour. It is medium sized, measuring 10-15 mm, with a wing expanse of 20-25 mm. The caterpillar feed hidden inside silk galleries on the underside of leaves.

Nature of Damage

As a result of the numerous galleries made by the feeding caterpillars the foliage dries up. Infested trees can be recognized from the dries up patches in the fronds.

Life Cycle

A female moth lays 125 scale like eggs in small batches on the underside of the tips of the old leaves. The incubation period lasts about 3-5 days in summer and 10 days in winter. The larvae is full-grown in 40 days and pupate inside the galleries. The pupal period lasts for 12 days. The adult moths survive for 2-3 days The pest has 7 generations in a year.

Management

1. Remove and destroy the infested leaves along with the caterpillar.
2. The infestation can be minimized by giving regular irrigation.
3. The coconut cultivar *viz.*, hybrid T x D, hybrid D x T, and west coast tall are tolerant to the pest attack.
4. To minimize the pest population in the coconut nursery spraying of endosulfan 0.07 per cent or monocrotophos 0.05 per cent is recommended.
5. Monocrotophos 36 per cent WSC @ 5 ml per tree (below 15 years aged tree) and @ 10 ml per tree (15 years and above age) with equal quantity of water applied by root application method has been recommended for effective control of this pest.
6. The larval parasitoid, *Goniozus nephantidis* can be utilized for the biological control of this pest. The pupal parasitoid, *Brachymeria nephantidis* is also proved to be more potential bioagents to suppress this pest.

Rhinoceros Beetle, *Oryctes rhinoceros* (Coleoptera: Curculionidae)

Marks of Identification

The adult is a stout-build black beetle, measured 35-50 mm in length, 14-21 mm in breadth and with a cephalic horn which is longer in male. The pygidium is densely clothed with reddish brown hairs in the female. The full grown grub is of stout, cylindrical but strongly arched convexly above and concavely beneath.

Nature of Damage

The adult beetle injures the trees by boring into the growing spear leaf cluster, spathes and petioles. The boring beetle chews the internal tissues and after imbibing the juicy part, throws out the fibrous part which protrudes out of the holes. The damaged leaf when emerges out is appeared like leaflets cut by scissor. The attacked central shoot topples down and repeated attack may even kill young palms. The damage to the spathes causes reduction in yield up to 10 per cent.

Life Cycle

A female may lay 100-150 eggs which hatch in 8-18 days and the grub start feeding on the decaying matter found in the vicinity. The grub pass through three instars to complete their development in 99-182 days, pupation takes place in chamber at a depth of about 30 cm and the beetle emerge after 10-20 days. They remain in the pupal cell for about 11-20 days before coming out of the soil and on emergence they are soft bodied creatures. The beetles are active at night and may be attracted to a source of light. The adults can live for more than 200 days. Generally, one generation is completed in a year.

Management

1. Avoid the manure pits on orchard or nearly the orchard to prevent the egg deposition and feeding of this pest.
2. Filling up the crowns, particularly the inner most two-three leaf axils with insecticide methyl parathion dust and sand mixture at least twice in a year before and after the rainy season is an excellent prophylactic measure against the beetles.
3. Rotting castor cake is recommended to attract and trap the adult beetles. For this water soaked castor cake treated with carbaryl 0.1 per cent on a w/ w basis is to be exposed in orchard in small mud pot.
4. Floor of manure pits/compost pits should be treated with contact insecticide (methyl parathion dust) to control the full grown grubs which reach the floor of the pits for pupation in the soil below.
5. Insecticidal treatment of the cattle manure or compost with methyl parathion 2 per cent dust should be given to control the immature stages of this beetle.
6. Mechanical control involves examing the crowns periodically and extracting the adult beetles by means of a hooked pointed metal rod about 0.5 m long with hook at one end, and fill up the hole with a mixture of methyl parathion 2 per cent dust and an equal volume of fine sand to prevent the reinfestation of this pest.

14
Pest of Spices

Chilli

Chilli Thrips, *Scirtothrips dorsalis* (Thysanoptera: Thripidae)

Marks of Identification

Insects have modified piercing sucking mouthparts. These insects are very small, about the size of flea. They are just visible to the naked eyes. The young are yellow or white. Adults are darker and brownish with or without strips on their back.

Nature of Damage

Both the nymphs and adults lacerate the tissues and suck the sap from tender leaves, growing shoots, flower buds and developing fruits. The leaves of affected plant curl and shed while the buds become brittle and drop down. They are also associated with *murda* disease.

Life Cycle

The pest is active throughout the year except during the rainy season. The female thrips lays 45-50 eggs inside the tissues of the leaves and shoots. The eggs hatch in 5 days. The larvae full-fed 7-8 days and pupate in 2-4 days. The adult live for 31 days. There are several overlapping generations in a year.

Management

1. Grow resistant/tolerant varieties *viz.*, Bonapari, LIC-13, 36, 45.
2. Parasitoids like *Orius maxidentex*, *Chrysoperla cornea*, *Amblyseius longispinosus*, *Hauptamannis* sp. and *Cheiracantius* sp. are key enemies of thrips, encourage their activities.

3. Spray monocrotophos (0.5 per cent) or methyl-o-demeton (0.025 per cent)
4. Soil application of carbofuran @ 0.5 kg a. i. per ha or phorate 0.7 kg a. i. per ha 15 days after transplanting control pest effectively.

Whitefly, *Bemisia tabaci* (Hemiptera: Aleyrodidae)

Marks of Identification

The louse-like nymphs clustered together on the under surface of the leaves and their pale yellow bodies make them stand out against the green background. In the winged stage, they are 1.0-1.5 mm long and their yellowish bodies are slightly dusted with a white waxy powder. They have two pairs of pure white wings and have prominent long hindwings.

Nature of Damage

Adults and nymphs suck the cell sap from lower surface of leaves and cause chlorotic yellow spots on upper surface of affected leaves. Whitefly also excretes honey dew, which make the leaves sticky. Sooty mould (*Cladosporium* Sp.) growth on such leaves interferes with photosynthesis of plants.

Life Cycle

A female can lay as many as 150-200 eggs. On an average, 28-43 eggs were laid singly on lower surface of leaves in a oviposition period for 2-18 days. Eggs are stalked, light yellow in colour and measure 0.2 mm. They hatch in 3-5 days. The nymphs on emergence feed on cell sap and grown in three stages to form the pupae within 9-14 days. The life cycle is completed in 14-122 days and many generations are completed in a year

Management

1. Whiteflies can be effectively attracted and controlled by yellow sticky traps, which are coated with grease or sticky oily material.
2. Use whitefly tolerant varieties such as LPS-141 (Kanchan), LK-861 and NA-1280.
3. Spray trizophos (2.5 ml per litre of water) or profanophos (2 ml per litre of water).

Chilli Fruit Borer, *Helicoverpa armigera* (Lepidoptera: Noctuidae)

Marks of Identification

The moth is stoutly build and is yellowish brown. There is a dark speck and a dark area near the outer margin of each forewing. The forewings are marked with grayish wavy lines and black spots of varying size on the upper side and a black kidney shaped mark and a round spot on the underside. The hindwings are whitish and lighter in colour with a broad blackish band along the outer margin. The caterpillar when full grown is 3.5 cm in length, being greenish with dark brown grey lines along the sides of the body.

Life Cycle

The female lay eggs singly on tender part of the plants. A single female may lay as many as 741 eggs in 4 days. They hatch in 2-6 days. The young larvae feed on the foliage for some time and later bore into the bolls and feed inside. They full-fed in 13-19 days. The full grown larvae come out of the boll and pupate the soil. The pupal period lasts 8-15 days. There may be as many as 8 generations in a year. The caterpillars feed on their fellows if suitable vegetation is not available.

Management

1. Grow simultaneously 40 days old American tall marigold and 25 days old tomato seedlings at 1:16 rows to attract the female moth for egg laying.
2. Set up pheromone traps with *heli* lure at 15 per ha and change the lure once in 15 days.
3. Collect and destroy the damaged fruits and grown-up caterpillars.
4. Six release of *Trichogramma chilonis* @ 50,000 per ha per week coinciding with flowering time and based on ETL.
5. Spray *Helicoverpa armigera* NPV at 500 LE per ha along with cotton seed oil (300 g per ha), to kill larvae.
6. Spray endosulfan 35 EC @2 ml per litre or *Bt.* @ 2 g per litre.
7. Encourage activity of parasitoids, *Eucelatoria bryani, Campolites, Chelonus*, etc.

Tobacco Caterpillar, *Spodoptera litura* (Lepidoptera: Noctuidae)

Marks of Identification

Adults are stout, with wavy white markings on the brown forewings and white hindwings having a brown patch along its margin. Full grown caterpillar is pale brown with greenish to violet tinge. There are yellow and purplish spots present in the sub-marginal areas.

Nature of Damage

Freshly hatched larvae feed gregariously, scrapping the leaves from ventral surface. They feed voraciously during night and hide in the morning. Entire crop is defoliated overnight.

Life Cycle

Female lays dirty white coloured eggs in cluster on the under surface of the leaves and covered with brown hair. Incubation period is 3-5 days. Larval period is 20-28 days. Pupation takes place in the soil earthen cocoon for 7-11 days. Total life cycle is completed in 30-40 days during summer and 120-140 days in winter season. There are 7-9 overlapping generations in a year.

Management

1. Plough the soil to expose and kill pupae.
2. Grow caster along border and irrigation channels as indicator or trap crop.
3. Flood the field to draw the hibernating larvae.
4. Set up light or pheromone traps as 15 per ha.
5. Remove and destroy egg mass in castor and tomato.
6. Collect damaged leaves and gregarious early instar larvae and destroy.
7. Hand pick grown-up larvae and kill them.
8. Spray NPV for *S. litura* 250 LE along with teepol 1 ml per ha in evening hours.
9. Spray chlorpyriphos 20 EC 2.0 litre per ha or DDVP 76 WSP 1.0 litre per ha or endosulfan 35 EC 1.25 litre per ha or NSKE 5 per cent.
10. Prepare poison bait with rice bran 5 kg, jaggery 0.5 per cent, carbaryl 50 WP 0.5 per cent, water 3 litre per ha and spread the bait in the evening hours.

Broad Mite, *Polyphagotarsonemus latus* (Acarina: Tarsonemidae)

Marks of Identification

Individuals are extremely small, about the size of a grain of sand and not clearly visible to the naked eye. They are found in groups hidden around the mid-vein on the undersides of the leaves. They appear crab like and are yellow or white.

Nature of Damage

Damage is usually confined to underside of leaves, where areas between veins are brownish and dried out and brittle in severe cases. Young leaves are cupped downward and narrower than normal.

Life Cycle

Female lay eggs singly on the surface of the leaves. These eggs are white, oval and extremely large compared to adults that lay them. Populations are continual but appear to be limited at high temperatures. This mite also fed on tomato, potato, beans and pepper. Only males are produced from eggs of unmated females.

Management

1. Application of sulphur @ 2.5 kg or dicofol 0.1 to 0.3 kg per ha controls this mite pest effectively.

Black Pepper

Pollu Beetle, *Longitarsus nigripennis* (Coleoptera: Chrysomelidae)

Marks of Identification

The adult is small shiny yellow and blue flea beetle with stout hind legs. The full grown grub is yellowish with a black head and it measures 5 mm in length.

Nature of Damage

The grub causes damage by boring into the berries and eating the contents completely in about 10-11 days. Each grub destroys at least 3-4 berries during the larval period. The attacked berries appear dark in colour, are hollow inside and crumble when pressed. The grubs may also eat into the spike and cause the entire distal region to dry up. The adult feed voraciously on tender leaves and make holes in them.

Life Cycle

The female makes holes on the berries and lay 1-2 eggs in each hole. A female on an average, lays about 100 eggs. The eggs hatch in 5-8 days and the young grubs bore into the berry and feed for 20-32 days. Then they drop to the ground and pupate in an earthen cell in the soil at 5-7.6 cm deep. The pupal period is of 6-7 days. The total life cycle completed in 30-50 days. There are four overlapping generations in a year.

Management

1. Tilling the soil in the base of vines at regular interval can reduce the population considerably.
2. Spray quinalphos (0.1 per cent) or dimethoate (0.03 per cent) in late July and again in early October.

Leaf Gall Thrips, *Liothrips karnyi* (Thysanoptera: Tripidae)

Marks of Identification

The adults are black and measure 2-3 mm in length. The larvae and pupae are creamy-white in colour.

Nature of Damage

Both nymphs and adults infest leaves of black paper and reduce the formation of tubular marginal leaf galls within which they live. The pest infestation is generally serious at higher altitudes and in nurseries in the plains. Apart from the formation of marginal leaf galls, the pest infestation results in reduction in size, crinkling and malformation of the infested leaves. In severe case of infestation, the growth of the vine and sometimes the formation of spikes may be adversely affected.

Life Cycle

The eggs are laid within the galls and they hatch in 6-8 days. The two larval stage, pre-pupal stage and pupal stage lasts for 4-7, 4-6, 2, 2-3 and 2-3 days, respectively.

Management

1. *Montandoniola moraguesi* and *Androthrips flavipes* are the common predators of this pest.
2. Spray monocrotophos (0.05 per cent) or dimethoate (0.05 per cent) for controlling the pest.

Top Shoot Borer, *Cydia hemidoxa* (Lepidoptera: Tortricidae)

Marks of Identification

Adults are small with a wing expanse of 10-15 mm, the forewings being crimson-red and yellow and the hindwings are grey. Fully grown larvae are grayish-green and measure 12-14 mm in length.

Nature of Damage

The larvae cause damage, the earlier instars of the larvae live within the silken webs on the tender shoots and scrape and feed on them. Later, the larvae bore into the tender shoots and feed on the internal contents resulting in drying up of the tender shoots. Repeated infestations of new shoots affect the growth of the vine.

Life Cycle

The female lay eggs on the tender portion of the plant. After hatching from eggs the earlier larvae live in silken webs on the tender leaves and feed them by scraping green matter. The larval period lasts for 14 days. Pupation generally occurs within the infested shoots and sometimes outside. The pupal period lasts for 8-10 days. This pest is more active in the field during August- December.

Management

1. The larval population is parasitized by hymenopteran parasite *viz.*, *Apanteles* sp., *Eudederus* sp. and *Goniozus* sp.
2. Spray endosulfan (0.05 per cent) or monocrotophos (0.05 per cent) during June and September is very effective in controlling the pest infestation.

Black Pepper Mussel Scale, *Lepidosaphes piperis* (Hemiptera: Diaspididae)

Marks of Identification

Full grown females are small dark brown and boat shaped measuring about 3-5 mm in length. The size of males is 2 mm.

Nature of Damage

Mussel scale encrust the main stem of young veins and lateral branches, mature plants, spikes, berries and petioles of pepper veins. The female scale sucks the sap causing serious injuries on feeding parts. Except for roots, all other parts have been terribly damaged. The infestation results in chlorotic spots, or patches, yellowing and necrosis of leaves due to desaping. Young veins succumb to the infestation whereas older infested lateral branches wilt and dry in patches resulting in necrotic vacant spaces in the canopy. Infestation by mussel scale causes significant loss of production as it affects all parts of plant including berries.

Life Cycle

The eggs are laid by the female under the scale cover and crawlers that hatch out from the eggs move about and fix themselves on the host plant at suitable places and thereafter become sedentary during the remaining part of their life. Early instar nymphs are spindle shaped. Once feeding is commenced white silken threads are being secreted and cover the posterior part of abdomen. After a period of 5 days white coloured silken threads become melanised and turn darker, and harder. The total life cycle lasts for about a month.

Management

1. Pruning and destroying severely affected lateral branches.
2. Number of parasitoids and predators (*Chilocoris circumdatus, Sarajiscymnus dwipakalpa, Pseudoscymnus* sp.) are effective in regulating the pest population so encourage their activities.
3. Apply neem formulations 0.5 per cent and fish oil insecticidal soap (3 per cent) four times at fortnight intervals after the harvest of berries suppress the pest very effectively.
4. Spray dimethoate (0.05 per cent) or monocrotophos (0.05 per cent) two times at 15 days interval control the pest effectively.

Ginger and Turmeric

Shoot Borer, *Conogethes punctiferalis* (Lepidoptera: Pyralididae)

Marks of Identification

The adult is a small moth with orange, yellow wings with small black spots. Full grown larvae are light brown and 16-26 mm in length.

Nature of Damage

The larvae bore into pseudostems and feed on the growing shoot resulting in yellowing and drying of infested shoots. The presence of bore holes in the pseudostem through which frass is extruded and the withered central shoots are the characteristic symptoms of pest infestation.

Life Cycle

Adult female laid 30-60 eggs during its life span. The egg period last for 3-4 days. There are five larval instars and they last for 3-4,5, 3-7, 3-8 and 7-14 days, respectively. The pre pupal and pupal period lasts for 3-4 and 9-10 days, respectively.

Management

1. Grow resistant/tolerant varieties of ginger- Rio-de-Janerio and turmeric, Dindigam Co-69 and Mannuthy local.
2. Spray malathion (0.1 per cent) during July- October is the most effective way to control the pest.

Leaf Roller, *Udaspes folus* (Lepidoptera: Hesperidae)

Marks of Identification

Adults are medium sized with brownish black wings having large white spots. The mature larva is dark green and is about 36 mm in length.

Life Cycle

The egg, larval and pupal periods lasts for 4-5, 13-25 and 6-7 days, respectively on ginger. On turmeric the various stages lasts for 3-4, 2-21 and 6-7 days, respectively.

Management

1. Spray dimethoate (0.05 per cent) for the control of the pest.

Rhizome Scale, *Aspidiella hartii* (Hemiptera: Diaspidae)

Marks of Identification

The adult females are minute, circular and light brown to grey.

Nature of Damage

The rhizome scale infests rhizomes of ginger and turmeric, both in field and in storage. In the field in severe cases of infestation, the plants wither and dry. In storage the pest infestation results in shriveling of buds and rhizomes. When the infestation is severe, it adversely affects sprouting of rhizomes.

Life Cycle

Females are ovo-viviparous and also reproduce parthenogenetically. About 100 eggs are laid by a single female.

Management

1. Dipping turmeric seed rhizomes in monocrotophos (0.1 per cent) for 5 minutes after harvest and before planting is effective in controlling this pest infestation.
2. The natural enemies *Physcus comperei* and *Adelencyrtus moderatus* are effective in regulating the pest population under the field conditions.

Cardamom

Cardamom Thrips, *Sciothrips cardamomi* (Thysanoptera: Thripidae)

Marks of Identification

The head and abdomen of this thrips are green brown and thorax and legs are pale yellowish brown. Adult measure approximately 1.2-1.5 mm in length.

Nature of Damage

Injury is caused by feeding in young leaf sheaths and basal ends of unripened flower bracts. Nymphs and adults feed with their piercing–sucking mouthparts, which lacerate young tissues and suck the juices that ooze from ruptured cells. First the injured area develops a silvery sheen because of air occupying the emptied cell cavities. Latter on this area become whitish, yellow, and brown splotches and streaks. Because of the secretive nature of feedings only within the sheaths and unopened bracts, its early presence usually goes undetected, until one actually looks down into the opened inflorescence and see the feeding damage at the bottom of the flower bracts. This thrips is suspected to be a vector of a mosaic disease in cardamom.

Life Cycle

The pest is active throughout the year except during the monsoon season. The female lays 5-71 eggs at random on all the feeding area of the plant. The young nymphs emerge from the eggs in 9-12 days. The first two nymphal instars are active and grow by feeding on the plant sap. Pupal period lasts for 10-15 days. Life cycle is completed in 25-30 days.

Management

1. Clean cultivation practices *viz.*, sanitation such as the removal of alternative hosts and plant debris within and outside the field.
2. Removal of dried shoot peels at the base of the stem resulted in a reduction of thrips population by one third.
3. Regularly spraying the cardomon crop either with fenthion 0.025 per cent or phenthoate 0.05 per cent or phosalone 0.07 per cent or methyl parathion 0.05 per cent or chlorpyriphos 0.05 per cent once in 3days during summer months and once in 45 days during the winter months.

Banana Aphid, *Pentalonia nigronervosa* (Hemiptera: Aphididae)

Marks of Identification

The wingless aphid is dark brown pyriform measuring 1.34 mm in length and with six segmented antennae which are longer than its body. Abdomen is dark brown, shining and slightly buldged. The winged forms are dark brown, elongated and pyriform. They are longer than the wingless form but body width is less.

Nature of Damage

The aphids feed on the leaf sheath and pseudostem. The insect cause little direct damage, but is of considerable significance being vector of cardamom mosaic in small cardamom.

Life Cycle

The reproduction takes place parthenogenetically. The longevity of adult varies from 8-26 days with an average of 14 days. A single female lays 8-28 off-springs with an average of 14. The development is completed through three to four moults in 12-15 days, respectively. There are 21-24 generations in a year.

Management

1. Spraying of dimethoate (0.05 per cent) is effective for the control of this thrips.

Castor Capsule Borer, *Dichocrosia punctiferalis* (Lepidoptera: Pyralidae)

Marks of Identification

The full-grown caterpillar measures 25-30 mm in length, is reddish brown with black blotches all over the body and pale strip on the lateral side. The moths are orange yellow, with black markings on both the wins.

Nature of Damage

This is a serious pest of nursery plants and young green pods of cardamom also. In the nursery and main plants, it bore into the stem and causes the death of the central core. It also eats away the tender seeds of the young berries.

Life Cycle

The female moths lay eggs on leaves and other soft parts of the plant. The eggs hatch in about a week. The larvae pass through 4-5 instars and are full-fed in 2-3 weeks. Pupation takes place inside the shoot or sometimes in the frass that collects after feedings. The pupal stage lasts about one week. The life cycle is completed in 4-5 weeks and there are about 3 generations in a year.

Management

1. It is advisable that the infested shoots and barriers may be collected and destroyed.
2. Spray quinalphos or methyl parathion (0.05 per cent) during early blooming period, it helps in reducing the pest damage.

Cardamom Hairy Caterpillar, *Lenodera vittata* (Lepidoptera: Arctiidae)

Marks of Identification

The adult moth is stout and fairly big and densely covered with scales. The larvae are clothed with a dense felt of capitate hairs and measure 106-110 mm in length.

Nature of Damage

The caterpillars are voracious feeders and cause capsule extensive damage to the cardamom plants by feeding on leaves. They are active from August to December. Only pseudostems and midribs remain un-eaten in case of severe infestation.

Life Cycle

The moth lay cream coloures dome shaped eggs in rows on both the upper and the lower surface of leaves. A single female lays 100-130 eggs during an oviposition period of 6-9 days. The young larvae emerge from the eggs in 10-13 days. The larvae start feeding on the leaves and other tender parts of the plant. They moult 6 times during the larval period of 110-120 days. Pupation takes place in the soil in earthen cell, which it stays for 5-7 months. There is only one generation in a year.

Management

1. The larvae is parasitized by a tachinid fly, *Carcelia kockrana*.
2. Spray 0.05 per cent malathion or 0.05 per cent monocrotophos to protect the crop from this pest.

Hairy Caterpillar, *Eupterote cardamomi* (Lepidoptera: Arctiidae)

Marks of Identification

The adults are large moths, ochreous in colour, with post-medial lines on the wings. They measure 70-80 mm in wing expanse. The larvae are hairy, dark grey in colour with pale brown head, bearing conical tufts of hairs on the dorsal side of the body. Full-grown caterpillar measures 90 mm in length.

Nature of Damage

The larvae feed on leaves of the shade trees up to the 6th or 7th instar and then they drop down on the cardamom plants growing underneath, with the help of silken threads. They start feeding on the leaves of cardamom voraciously and defoliate the cardamom plants causing heavy reduction in the yield.

Life Cycle

The moths emerge with start of monsoon rains in the month of June and July. The female moths lay 400-500, yellowish and dome-shaped eggs in the flat masses on the undersurface of the leaves. Each mass contains about 50-60 eggs. The hatching occurs in 15-17 days. The larva passes through 10 instars in 140-150 days. It pupates in the soil in a silken cocoon at a depth of 5-8 cm. The pupal period lasts for 7-8 months. The adult moths live for about 20 days. There is only one generation in the year.

Management

1. Larvae are parasitized by *Stuemia sericariae* and *Apantelis eupterote* under the field condition.
2. Spray 0.05 per cent malathion or 0.03 per cent endosulfan for its management.

Rhizome Weevil, *Prodioctes haematicus* (Coleoptera: Curculionidae)

Marks of Identification

The adult are a brown coloured weevil measuring 12 mm in length.

Nature of Damage

The grub causes tunnels in the rhizome by feeding results in the death of entire clumps of the cardamom plant.

Life Cycle

The female weevil lays egg in cavities made on rhizomes. The young grubs come out of the eggs in 8-10 days and bore into the rhizome, making tunnels. They feed inside the rhizome and full-fed in 3 weeks. They pupate within the feeding tunnels. The pupal period lasts for 3 weeks. The adult weevil lives for 7-8 months. There is only one generation in a year.

Management

1. Destroy affected plants and seedlings.
2. If the grub population is more in soil, drench the base of the clump with 1.25 litre of malathion.

Cardamom Root Grub, *Basilepta fulvicorne* (Coleoptera: Scarabaeidae)

Marks of Identification

The adults are shiny metallic blue, green brown beetles measuring 4-6 mm. Males are smaller than the female. The fully grown grubs are pale white, stout and 'C' shaped.

Nature of Damage

Root grubs are major pest of cardamom in nurseries and main field. The pest is generally serious in primary and secondary nurseries. The grubs feed and damage roots and portions of rhizomes, sometimes the entire root system is eaten away. The infested plants turn yellow and are stunted, severely infested plants dies.

Life Cycle

The female lays eggs in dried leaves around the base of clumps. The egg, larval and pupal stages lasted for 3-19, 45-60 and 10-17 days, respectively.

Management

1. Collection and killing of adult beetles during March- April and August-September during peak period of emergence.
2. Apply chlorpyriphos (0.04 per cent) or phorate 10 G (20-40 g per clump) during May and October.

Coriander

Coriander Aphid, *Hyadaphis coriandari* (Hemiptera: Aphididae)

Marks of Identification

Nymphs and adults are pear shaped, light green in colour, their blue white look is due to powdery substances present on the body. Nymphs are 11 mm and adults are 12 mm long.

Nature of Damage

Both nymphs and adults are found to suck the sap from leaves, flowers and immature seeds mainly attacking umbels. Beside this they also excrete honeydew, which favours the growth of shooty mould, as a result the growth of plant retards, and quality and quantity of seeds are also affected. In case of severe infestation at early stage of development of tender leaves, growing points, flower stalks withers and dries up. When the aphid infestation occurs at flowering and seed seeting stage, the seeds are not formed and if they formed they are shriveled and of poor quality. In case of severe infestation there is complete failure of crop.

Life Cycle

The aphid breed throughout the year by parthenogenesis and viviparous reproduction. Sexual reproduction takes place during severe winter season. A single female in her life produces about 40-50 nymphs. All young ones are apterous and viviparous females. The winged forms are produced when the population increase very sudden overcrowding, partial starvation, high temperature and low relative humidity conditions occurred. The nymphs moult three times and nymphal period lasts for 8-10 days. The adult period is for 3-4 days and total life cycle normally completes in 14-21 days during summers and as long as 35-40 days in winters.

Management

1. Timely sowing of coriander play an important role in occurrence of the aphid. If time of sowing is manipulated, the attack of this aphid will automatically reduced.
2. Lady bird beetles *Coccinella septumpunctata* and *Menochilus sexmaculatus* are very important predators of this aphid under field condition, so always conserve and protect them.
3. Aphid incidence can be controlled by two sprays with monocrotophos or quinalphos or endosulfan (0.1 per cent) at 10-15 days interval *i.e.*, at the time of floral initiation and 15 days thereafter.

Green Peach Aphid, *Myzus persicae* (Hemiptera: Aphididae)

Nature of Damage

This is a very serious pest of coriander. Damage of *M. persicae* is confined to the umbels only. Nymphs and adults suck the sap from the umbels. Due to honeydew secretion from them sooty moulds are also developed on umbels which results in poor development of seed.

Life Cycle

The female produces 5-15 ovipare. These are fertilizing by the male and each ovipare then lays 4-14 eggs. The eggs undergo during winter. Hatching of eggs coincides with vegetative growth of the plant. Newly hatched young once feed on soft portions of the plants and develops rapidly and start reproducing parthenogenetically.

Management

1. Yellow sticky traps attract the winged aphids and can be utilized in trapping the pest. For this purpose number of materials like yellow polythylex, sticky traps and water traps have been used in minimizing the population of this pest.
2. Early sown crops upto the middle of November has low infestation then the crop sown after this.
3. Natural enemies like Coccinellid predators *viz., Coccinella septumpunctata, Bromoides suturalis, Menochilus sexmacalatus* and *Adonis* sp. have been common predators of this aphid.
4. The pest is controlled by spraying endosulfan (0.07 per cent) twice just at the time of floral initiation and 15 days thereafter.

Cotton Whitefly, *Bemisia tabaci* (Hemiptera: Aleyrodidae)

Marks of Identification

The louse like nymphs, which suck the sap are sluggish creatures, clustered together on the under surface of the leaves and their pale-yellow bodies make them stand out against the green background. In adult stage, they are 1.0-1.5 mm long and their yellowish bodies are slightly dusted with a white waxy powder. They have two pairs of pure white wings and have prominent long hindwings.

Nature of Damage

Nymphs and adults both cause damage to the crop. They remain underside the leaves and suck the sap. Continuous sucking of sap from the leaves results in chlorotic spots and later on, the leaves coalesce and become brittle and finally dropdown from the plant prematurely. The honeydew secreted by the pest drop on the upper surface of the lower leaves and helps in development of shooty moulds which interferes with photosynthesis of leaves.

Life Cycle

the female generally lay eggs on the under surface of the leaves. The eggs are inserted in the leaf tissues. A single female lays 200-300 eggs either singly or in groups, which are yellowish in colour and turn brown to dark before hatching. Incubation period is 3-5 days. There are 4 nymphal instars, fourth nymphal instar is called pupa. Generally, nymphal period is 9-14 days. The pupal period lasts for 2-8 days. This pest generally completes its life cycle in about 13-20 days. There are 11-13 generations in a year.

Management

1. Judicious use of nitrogenous fertilizers and irrigation as well as proper spacing in plants can help in check of whitefly population build up.
2. Use of fish oil soap at 2 per cent, neem oil (0.5 per cent) or methyl-o-demeton (1 litre per ha) for suppressing the pest population.

Coriander Cutworm, *Spodoptera exiguva* (Lepidoptera: Noctuidae)

Marks of Identification

The moths have dark spotted forewings and white hindwings. They are active at night but remain hidden under various shrubs in the day time. The colour of larvae is light green and they are 30 mm long.

Nature of Damage

The larvae cause damage to the crop. The damaged crop, on which larvae have fed, gives a webbed appearance. The older caterpillars, which feed in morning and evening, have voracious feeder. They feed large amount of leaves.

Life Cycle

A female lays upto 600 eggs in clusters. The eggs are spherical and resembles poppy seed in shape and size, having lines radiating from the center. The egg clusters are covered with buff hairs. The eggs hatch in 1-3 days and the young caterpillars start feeding in groups. The larvae become full-fed in 15-20 days. Pupation occurs in soil. The pupal stage lasts for 5-7 days and the life cycle is completed in abut 30 days. There are 8-10 generations in a year.

Management

1. Activity of this pest can be suppressed by keeping field clean.
2. The pest can be suppressed by collecting and destroying the egg masses.
3. *Euplectus* sp., is closely associated natural enemy of this pest.
4. Spray endosulfan (0.07 per cent) to suppress the pest.

Flower Stink Bug, *Agonoseelis nubile* (Hemiptera: Pentatomidae)

Marks of Identification

The pest is about 12 mm long and its body colour is green or dark green with black spots. The scutelum is quite long and conical in shape.

Nature of Damage

Both nymph and adults cause damage to plants. They suck sap from leaves, flowers and unmatured seeds. Due to heavy feeding chlorotic spots developed on leaves. In case of heavy infestation the flowers drop prematurely from the plants and the grains developed from these flowers become shriveled.

Life Cycle

A very less information is available regarding the life cycle of this pest. The female lays its eggs in the stems of plant, and this is active from February to October.

Management

1. Hand picking of insect in case of small field and kitchen gardens.
2. Spray malathion (0.04 per cent) or endosulfan (0.05 per cent) for its effective management.

Green Stink Bug, *Nezara viridula* (Hemiptera: Pentatomidae)

Marks of Identification

Green stink bug are bigger in shape and size in comparison to flower stink bug. Its body is 15mm long and 8 mm broad. It is deep green coloured insect, whose scutelum is pointed.

Nature of Damage

Both nymphs and adults cause damage by sucking sap from stem, leaves, flower and green seeds. Due to there infestation the plant growth is badly affected, the flowers drop prematurely and the seed yield reduced.

Life Cycle

The female lays about 300 eggs in cluster on the surface of leaves. Eggs are whitish or yellowish in colour when freshly laid but turn to pink before hatching. Eggs are barrel shaped and 1 mm long. The incubation period is of about a week.

Nymphs are about 5-8 mm in length with beautiful colour patterns. They are brownish red in colour with multicolor spots. Freshly hatched nymphs remain in clusters around the egg raft and it is only after first moult disperse and start active feeding. The nymphal development takes about one month. There are four to five generation in a year.

Management

1. Hand picking of nymphs and adult bugs.
2. Spray dimethoate (0.03 per cent) or quinalphos (0.05 per cent) for effective control of this pest.

Cumin

Aphid, *Hyaloptenus arundini* (Hemiptera: Aphididae)

Marks of Identification

The adult and nymphs are lice like in shape and size and their body colour is deep green. Their colony generally found on the soft portion of the plant.

Nature of Damage

Both nymphs and adults cause damage to the crop. They suck the sap from the soft portion of the plant. In early growth stage due to their attack the plant growth is badly affected, while in letter stage the pest cause great reduction in the yield. Beside this the pest excretion honeydew on the leaves which attract the black mould, which again interferes on photosynthetic activities of the plant and ultimately cause reduction inn the crop yield.

Life Cycle

This aphid breeds throughout the year by parthenogenesis and viviparous reproduction. Sexual reproduction takes place during severe winter season. A single female in her life produces about 40-50 nymphs. All young once are apterous and viviparous females. The winged forms are produced when the population increases very sudden over crowding, partial starvation, high temperature and low relative humidity conditions occurred. The nymph moult 3 times and nymphal period last for 8-10 days. The adult period is of 3-4 days and the total life cycle normally completes in 14-21 days during summer and as long as 35-40 days in winter.

Management

1. Always grow aphid resistant varieties of cumin *viz.*, VC-187, VC-154, VC-822 and VC- 33.
2. Spray dimethoate (0.03 per cent) for effective control.

Fenugreek

Aphid, *Aphis craccivora* (Hemiptera: Aphididae)

Marks of Identification

Both winged and wingless forms are present in this aphid. The mature adults measure 1.75–1.90 mm length. After first moult nymph become yellowish green in colour and become 0.01–0.04 mm long. As the nymph mature its colour becomes darker. The mature adult becomes olive grey or dull grey in colour.

Nature of Damage

The nymph and adults suck the sap, usually from the underside of leaves. Infestation in early stages causes stunting of the plants as well as reducing their vigour. When the attack occurs at the time of flowering and pod formation, there is significant reduction in the yield. Due to honey dew secretion black mould developed on the leaves which interfere in photo synthesis activities.

Life Cycle

The offspring of the winged form may be wingless. Even without fertilization, the female may produce 8-20 young ones in a life span of 10-12 days. The young nymphs pass through four moults to become adult in 5-8days. The apterous female start producing broods within 24hours of attaining that stage. Breeding occurs almost throughout the year, and there are many overlapping generations in a year.

Management

1. Natural enemies like Coccinellid predators *viz., Coccinella septumpunctata, Bromoides suturalis, Menochilus sexmacalatus* and *Adonis* sp. have been common predators of this aphid.
2. Spray endosulfan (0.03 per cent) or malathion (0.04 per cent) to protect the crop from this pest.

Flea Beetle, *Phyllotreta cruciferae* (Coleoptera: Chrysomelidae).

Marks of Identification

Adult beetles are elongate oval in shape and metallic bluish green in colour. Head has impunctate vertex and black antennae.

Nature of Damage

The adults mostly feed on the leaves by making innumerable round holes in the host plants. The affected leaves dry up. A special kind of decaying odour is emitted by the cabbage plants attacked by this pest.

Life Cycle

A single female lays 50-80 eggs in soil. Incubation period last for 5-10 days. The grubs feed on roots and do not cause much damage. Grub period lasts for 9-15 days. Pupal period is of 8-14 days. There are 7-8 generations in a year.

Management

1. The adults are parasitized by *Microctonus indicus.*
2. Spray quinalphos or chlorpyriphos (0.05 per cent) for effective control of this pest.

Fenugreek Leaf Minor, *Chromtomiya horticola* (Diptera: Agromyzidae)

Marks of Identification

The adults are two-winged flies having grayish black mesonotum and yellow frons. It is 2 to 3 mm long. The fully grown maggots are 3 mm long and 0.75 mm broad.

Nature of Damage

The larvae makes tunnels on the leaves interfere with photosynthesis and affect the growth of the plant, making them look unattractive.

Life Cycle

The female lay eggs singly in leaf tissues. The eggs hatch in 2-3 days and larvae feed between lower and upper epidermis by making zig-zag tunnels. They are full grown in about 5 days and pupate within the galleries. The adult emerges from the pupae in 6 days. The life cycle is completed in 13-14 days and there are many generations in a year.

Management

1. The attack of this pest is much more in drought condition so timely irrigation to the crop is very important.
2. Destroyed damaged leaves in its initial stage.
3. Chalcid, *Solenotus guptai* and *Rhopalotus thokerie* are the common parasitoid of maggots.
4. Spray dimethoate (0.03 per cent) or methyl-O-demetom (0.05 per cent) to protect the crop from this pest.

Leaf Webber, *Hymenia recurvalis* (Lepidoptera: Pyralidae)

Marks of Identification

Adult are very tiny moth, measuring 1.0 mm in length and 1.5–2.0 mm with wing expansion. The wings are dark green in colour with whitish spots. The larvae are dark green in colour with dark brown head.

Nature of Damage

The larvae bind leaves together with silken threads and then feed inside these leaves by scraping the green material. The infested leaves are not useful for consumption. In case of heavy infestation the leaves become dry and flowers, pods and seed formation do not takes place.

Life Cycle

Female lays 150 eggs in veins of leaves. Incubation period range between 3-4 days and the newly emerged larvae are dark green in colour. They moult 5 times to become fully mature. The fully mature larvae pupate inside soil in a cocoon. Pupal period lasts for 8-11 days. Adult live for 3-5 days.

Management

1. Pluck the affected leaves to reduce the larval infestation.
2. The adult moth attract towards light so use light trap to destroy the adult moths.
3. Braconid, *Apantelis delhensis* and *Cardiachiles* are larval parasitoids of leaf webber.
4. Spray malathion (0.5 per cent) for control of this pest.

Cinnamon

Cinnamon butterfly, *Chilasa clytie* (Lepidoptera: Lycaenidae)

Marks of Identificationz

The adults are large, males have bleakish brown wings with white spots on the outer markings, females have black wings with bluish white markings. Fully grown larvae are pale yellow with dark strips on the sides and are 25 mm in length.

Nature of Damage

This is the most destructive pest of cinnamon. The pest is generally seen in the field during December-June. The larvae feed on tender and slightly mature leaves with portions of veins reason.

Life Cycle

Eggs are laid on tender leaves and shoots and they hatch in 3-4 days. The larval stage comprising 5 instars and is completed in 11-17 days. The pupal period lasts for 11-13 days.

Management

1. The eggs are parasitized by *Telenomus remus*.
2. Spraying of quinalphos (0.05 per cent) for controlling the pest infestation.

15
Pest of Plantation Crops

Coconut

Leaf Eating Caterpillar, *Opisina arenosella* (Lepidoptera: Xyloryctidae)

Marks of Identification

The moth is ash grey in colour. It is medium sized, measuring 10-15 mm, with a wing expanse of 20-25 mm. The caterpillar feed hidden inside silk galleries on the underside of leaves.

Nature of Damage

As a result of the numerous galleries made by the feeding caterpillars the foliage dries up. Infested trees can be recognized from the dried up patches in the fronds.

Life Cycle

A female moth lays 125 scale like eggs in small batches on the underside of the tips of the old leaves. The incubation period lasts about 3-5 days in summer and 10 days in winter. The larvae is full-grown in 40 days and pupate inside the galleries. The pupal period lasts for 12 days. The adult moths survive for 2-3 days The pest has 7 generations in a year.

Management

1. Remove and destroy the infested leaves along with the caterpillar.
2. The infestation can be minimized by giving regular irrigation.
3. The coconut cultivar *viz.*, hybrid T x D, hybrid D x T, and west coast tall are tolerant to the pest attack.
3. To minimize the pest population in the coconut nursery spraying of endosulfan 0.07 per cent or monocrotophos 0.05 per cent is recommended.

4. Monocrotophos 36 per cent WSC @ 5 ml per tree (below 15 years aged tree) and @ 10 ml per tree (15 years and above age) with equal quantity of water applied by root application method has been recommended for effective control of this pest.
5. The larval parasitoid, *Goniozus nephantidis* can be utilized for the biological control of this pest. The pupal parasitoid, *Brachymeria nephantidis* is also proved to be more potential bioagents to suppress this pest.

Rhinoceros Beetle, *Oryctes rhinoceros* (Coleoptera: Scarabaeidae)

Marks of Identification

The adult is a stout-build black beetle, measured 35-50 mm in length, 14-21 mm in breadth and with a cephalic horn which is longer in male. The pygidium is densely clothed with reddish brown hairs in the female. The full grown grub is of stout, cylindrical but strongly arched convexly above and concavely beneath.

Nature of Damage

The adult beetle injures the trees by boring into the growing spear leaf cluster, spathes and petioles. The boring beetle chews the internal tissues and after imbibing the juicy part, throws out the fibrous part which protrudes out of the holes. The damaged leaf when emerges out is appeared like leaflets cut by scissor. The attacked central shoot topples down and repeated attack may even kill young palms. The damage to the spathes causes reduction in yield up to 10 per cent.

Life Cycle

A female may lay 100-150 eggs which hatch in 8-18 days and the grub start feeding on the decaying matter found in the vicinity. The grub pass through three

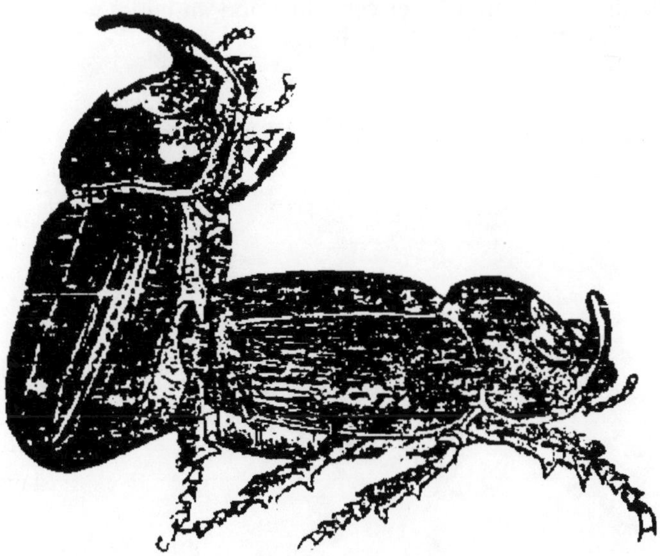

Figure 15.1: Rhinoceros Beetle

instars to complete their development in 99-182 days, pupation takes place in chamber at a depth of about 30 cm and the beetle emerge after 10-20 days. They remain in the pupal cell for about 11-20 days before coming out of the soil and on emergence they are soft bodied creatures. The beetles are active at night and may be attracted to a source of light. The adults can live for more than 200 days. Generally one generation is completed in a year.

Management

1. Avoid the manure pits on orchard or nearly the orchard to prevent the egg deposition and feeding of this pest.
2. Filling up the crowns, particularly the inner most two-three leaf axils with insecticide *viz.*, methyl parathion dust and sand mixture at least twice in a year before and after the rainy season is an excellent prophylactic measure against the beetles.
3. Rotting castor cake is recommended to attract and trap the adult beetles. For this water soaked castor cake and treated with carbaryl 0.1 per cent on a w/w basis is to be exposed in orchard in small mud pot.
4. Floor of manure pits/compost pits should be treated with contact insecticide (methyl parathion dust) to control the full grown grubs which reach the floor of the pits for pupation in the soil below.
5. Insecticidal treatment of the cattle manure or compost with methyl parathion 2 per cent dust should be given to control the immature stages breeding of this beetle in such media.
6. Mechanical control involves examing the crowns periodically and extracting the adult beetles by means of a hooked pointed metal rod about 0.5 m long with hook at one end, and fill up the hole with a mixture of methyl parathion 2 per cent dust and an equal volume of fine sand to prevent the reinfestation of this pest.

16
Stored Grain Pest

Rice Weevil (*Sitophilus* species)

Order–Coleoptera, Family–Curculionidae

Marks of Identification

They have a characteristic snout or rostrum which projects from the front of the head. The beetles biting mouthparts are located at the tip of the rostrum, and a pair of elbowed, clubbed antennae are located at the base.

Nature of Damage

Damage is done by adult and larvae. Grains are hollowed out. Kernels are reduced to mere powder. The pest infests grains in field too. Adults cut circular holes. Heating takes place during heavy infestation, which is known as dry heating.

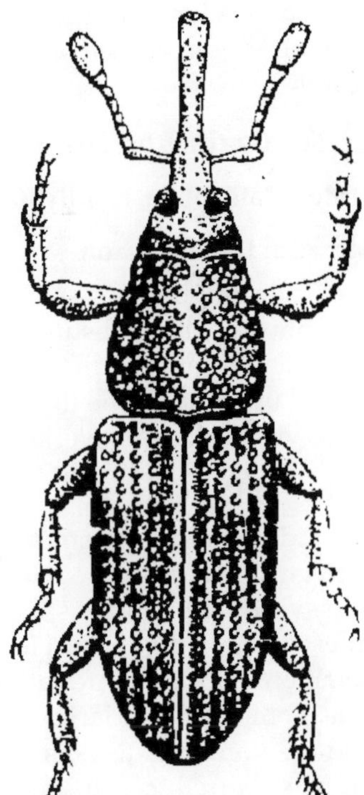

Figure 16.1: Rice Weevil

The three major *Sitophilus* species that are significant pests of stored grain are *S. oryzae* the rice weevil, *S. zeamais* the maize weevil, and *S.granarius* the granary weevil. In appearance *S. oryzae* and *S. zeamais* are identical with distinct yellow to orange coloured spots on their wing covers. *S. granarius* lack spots on its elytra and is a rich chocolate brown colour. Adult beetles are 2.5 to 3 mm long. *S. granarius* tends to be the larger of the three species. These three beetles are major pests of stored grain while *S. oryzae* and S. *granarius* may be found attacking all major cereals–*S. zeamais* is most often recorded in maize but is capable of infesting and developing on other cereals.

Biology

Eggs

The adult female bores a hole in the grain surface using its biting mouthparts. One egg is laid in this hole which is then plugged with a jelly substance sealing the egg inside the grain. The number of eggs laid by a female during its lifespan may vary from 100 to 450.

Larvae

After hatching the small legless larva feeds on the endosperm of the grain. Larvae are relatively immobile and pass through a number of instar stages gradually increasing in size. Larval stage lasts for about 20 days.

Pupae

When fully grown the larva passes into an inactive pupal stage and gradually assumes the shape and form of the adult. Pupal period is for 3-6 days

Adult

Adult is long lived live for 3-5 months and active feeder and are strong fliers. Infestations can cause considerable grain heating. There are 6-8 generations in a year.

Lesser Grain Borer (*Rhizopertha dominica*)

Order–Coleoptera, Family–Bostrychiade

Marks of Identification

The adult beetle is small, between 2.5 to 3 mm long, has a distinctive cylindrical shape, and is dark brown to black. The head, tucked underneath the thorax, is invisible when viewed from above.

Nature of Damage

Adult comes out from the grain leaving a irregular hole. In bagged storage irregular messy waste flour spots indicates infestation of this pest. Heating is very common. Localized infestation is almost a rule. Both adult and larvae cause damage and are voracious feeders. As such grain kernels are reduced to mere shells. The damaged kernels remain engulfed in a film of waste flour.

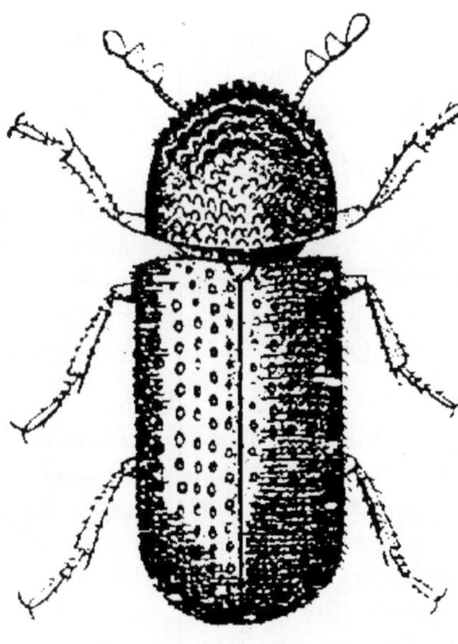

Figure 16.2: Lesser Grain Borer

Biology

Eggs

Laid in clusters as females actively bore through grains. Eggs are laid outside the kernel and young larvae bore into the grain to complete their development. Under optimum conditions the female lays up to 500 eggs during its lifespan. Eggs laid on stored commodities at moisture levels as low as 8 per cent can still hatch.

Larvae

The cream coloured larvae have biting mouthparts and three pairs of legs. The young larvae are mobile but become immobile as they complete their development concealed within grain or flour. The larvae normally pass through four instars during which their size increases. All larvae have usually bored into grain (or a suitable hard substrate) by the third instar.

Pupae

The mature fourth instar enters into an inactive pupal stage within the grain and gradually assumes the form of the adult.

Adult

When the pupal stage is completed the newly formed adult emerges from the grain by chewing through the outer grain layers. Entire life cycle takes 25 days under optimum condition. The adult beetle is long lived and is a strong flier when conditions are warm. They are adept burrowers and produce large quantities of flour. They are also capable of chewing their way through many types of packaging materials including jute, waxed paper and some polyethylene films. There are 6-7 generations in a year.

Saw-toothed Grain Beetle (*Oryzaephilus surinamensis*)

Order–Coleoptera, Family–Sylvanidae

Marks of Identification

The adult beetle is 2.5 to 3 mm long and can be readily identified from other beetle pests of stored products by the distinctive shape of its thorax–the section between the head and the body. It has six jagged, saw-toothed projections along the outer edges of the thorax. They are very active and their habit of infesting cracks and crevices makes them difficult to detect when inspecting warehouses their presence.

Nature of Damage

Adult and larvae cause roughening of grain surface and off odour in grain. Grains with higher percentage of brokens, dockage and foreign matter sustain heavy infestation, which leads to heating of grain.

Biology

Eggs

Laid at loosely in crevices in the commodity. An average 375 eggs are laid during the life of an adult female. Incubation period is for 3-17 days, limited to 5 days at higher temperature.

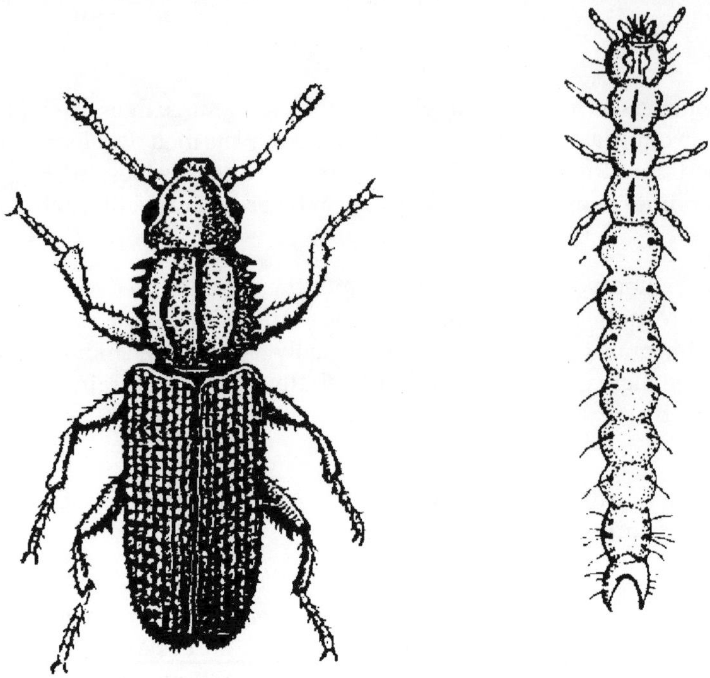

Figure 16.3: Saw-toothed Grain Beetle

Larvae

The larvae pass through four instars during their development. They are tolerant of dry conditions and development has been recorded at relative humidity as low as 10 per cent. Larvae are mobile and not concealed. Larval period last for 14-20 days.

Pupa

The pupal stage is formed within a cocoon spun by the mature larvae, and broken grains or other food particles may be used in its construction. Pre-pupal and pupal period last for 7-21 days.

Adult

Long lived live upto 3 years, feeds, flies, and will rapidly walk long distances. They can easily enter packaged food, and prefer to live in cracks and crevices. Its multiplication is quick in rainy season and coastal areas.

Flour Beetles

Red Rust Flour Beetle (*Tribolium casteneum*)

Confused Flour Beetle (*T. confusum*)

Order–Coleoptera, Family–Tenebrionidae

Marks of Identification

The rust-red flour beetle is between 3 to 4 mm long, flattened, reddish brown in colour and parallel sided. Eyes are crescent-shaped. The confused flour beetle is slightly larger, 4 to 4.5 mm long. Although similar in appearance, the confused flour beetle does not have a distinct club formed by the last three segments of each antenna, it has a distinct ridge above each eye, and the eyes are set further apart when viewed from underneath. Larvae are elongate, light brown. The beetle is a major pest of stored products, especially grain and milled cereal products.

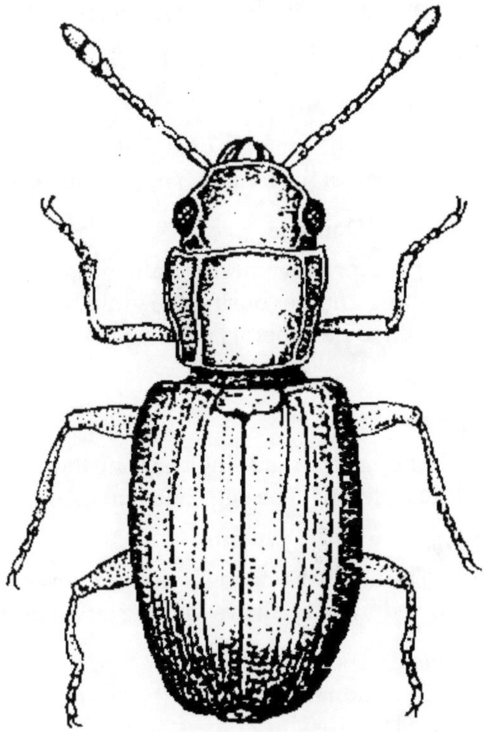

Figure 16.4: Red Rust Flour Beetle

Nature of Damage

Both adults and larvae feed on milled products. Floor beetles are secondary pests of all grains and primary pests of flour and other milled products. In grains, embryo or germ portion is preferred.The confused flour beetle prefers more finely divided commodities than the rust-red flour beetle.

Biology

Eggs

Laid at random in the commodity and more than 1,000 eggs may be produced by a female during its lifespan. Incubation period is 5-12 days depending upon temperature.

Larvae

Light brown colour, found free living and mobile in stored products. The larva has a pair of legs on each of the three segments immediately behind the head. Their size varies widely with age. The larva passes through several stages or instars during which it's size increases. The larval period is 3-12 weeks which is influenced by environmental conditions.

Pupae

When fully grown the larva enters into an active pupal stage and gradually changes into the form of the adult beetle. Pupal period last 5-9 days.

Adult

The newly emerged adults of both species are a creamy white colour but quickly darken to their typical red-brown to brown colour. Adults are long lived, actively feed, and heavy infestations often leave an unacceptable taint. Life cycle is completed in 4-5 weeeks. There are many generations in a year.

Khapra Beetle (*Trogoderma granarium*)

Order–Coleoptera, Family–Dermistidae

Marks of Identification

The beetle is small, oval, pale-red brown or black in colour. The male is almost half the size of the female. The antennae are small and clavate type.

Nature of Damage

It is a primary pest it damages the grain starting with germ portion, surface scratching and devouring the grain. Actually it reduces grain into frass. Damage is confined to peripheral layer of bags or 30-45 cm in bulk storage.

Biology

Eggs

The eggs are laid singly among the grains. The eggs are round, semi-transparent, white in colour. The eggs hatch in 6-16 days.

Larvae

The newly hatched larvae is yellowish white in colour letter on turns brownish followed by reddish brown. The larval period is 19-37 days. They moults 4 times.

Pupae

Pupation takes place in the last larval skin among the grains or in cocoon. Pupal period lasts for 4 to 27 days.

Adult

the adult beetles may survive for 25 to 35 days. The life cycle is completed under favourable condition in 34 to 76 days. There are 22-24 generations in a year.

Pulse Beetle (*Callosobruchus* spp.)

Order–Coleoptera, Family–Bruchidae

Marks of Identification

The adult beetles are small, squat with long conspicuous serrate antennae, colour is brownish grey with characteristic elevated ivory like spots near the middle of the dorsal side. Elytra do not cover the abdomen completely.

Nature of Damage

Larvae eat up the grain kernel and make a cavity. Adults are harmless and so not feed on stored produce at all.

Biology

Eggs

Female lays 80-100 eggs laid singly, glued to the surface of the grain and pods (in field). Incubation period is 5 days.

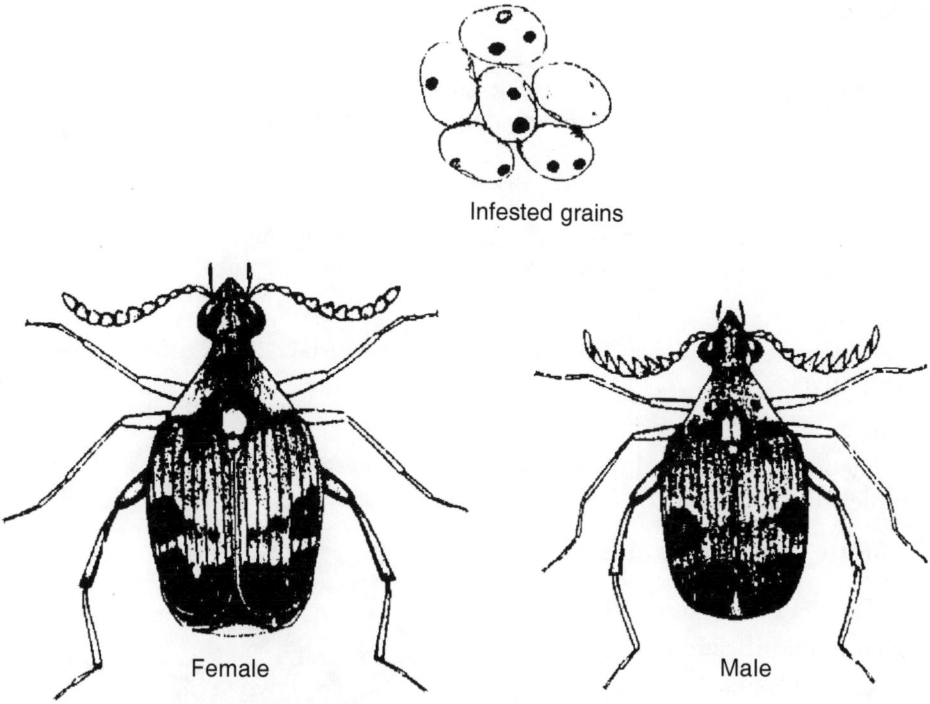

Figure 16.5: Pulse Beetle

Larvae

On hatching, bore into grain, making round, translucent 'windows' in seed before pupation. Larval stage lasts for 15-50 days

Adult

Emerges through 'window' leaving neat round hole. Short lived, does not feed on seed, runs quickly and flies very well. The life cycle completes in 23 days. There are 7-8 generations in a year.

Long-headed Flour Beetle (*Latheticus oryzae*)

Order–Coleoptera, Family–Tenebrionidae

Marks of Identification

The adult beetle is relatively small (2.5 to 3 mm), slender, oval shaped, and has a distinctive yellowish-brown colour. The head protrudes well past the front the eyes, and the antennae are short and thickened, with a five segmented clubs.

Nature of Damage

Secondary pest of foodgrains and is important pest of milled products. Both adults and larvae feed on stored products.

Biology

Eggs

The female adult lays about 300 eggs singly during her life span. The eggs which are sticky when laid are usually covered in flour particles. Incubation period is 7-12 days.

Larvae

The free living cylindrical larvae are white in colour and pass through 6 to 7 instars. Larval period is 15-80 days depending upon temperature.

Pupae

The pupae are white and are found naked amongst the foodstuff. Pupal period is for 5-10 days.

Adults

The slow moving adult is long-lived surviving up to 6 months. Life cycle completed in 25 days at 35 °C and 70 per cent R.H.

Cigarette Beetle (*Lasioderma serricorne*)

Order–Coleoptera, Family–Anobiidae

Marks of Identification

A small beetle 2 to 2.5 mm long, oval in shape and brown to dark brown in colour. It resembles the drug store beetle (*Stegobium paniceum*) however segments of each antenna are uniform in length and shape and the wing covers or elytra are not striated. The larvae is white in colour and short and stout in shape.

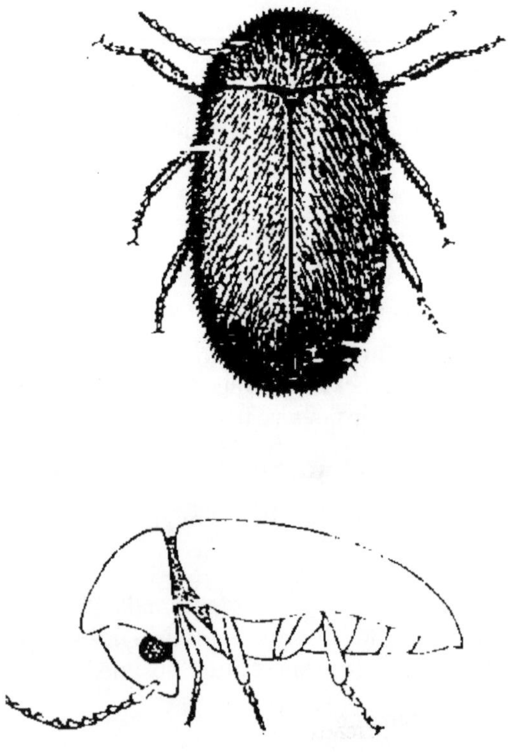

Figure 16.6: Cigarette Beetle

Nature of Damage

The tobacco or cigarette beetle attacks a wide range of stored products throughout tropical and sub-tropical climate and is particularly damaging to stored tobacco, cigars, cigarettes and cocoa beans.

Biology

Eggs

The eggs are laid singly in crevices or folds in the substrate. The adult female produces about 100 eggs in its short life-span of 25 to 30 days.

Larvae

The white, fleshy larvae are distinctly hairy and pass through 4 to 6 instars. Newly hatched larvae cannot attack whole grain. The larvae are mobile and readily enter small opening of packaged foodstuffs in search of food.

Pupae

When mature, the larve pupate within a thin cocoon built amongst the substrate. The adult spends some days within the cocoon before emerging from and boring through packaged materials.

Adult

In daylight, the adult beetle shuns light hiding in cracks and crevices. In warm conditions it flies readily at dusk and is attracted to artificial light. The species is not cold-hardy and can over winter in temperate climate.

Drug Store Beetle (*Stegobium paniceum*)

Order–Coleoptera, Family–Anobiidae

Marks of Identification

A small beetle 2 to 3.5 mm long, cylindical and light brown in colour. It resembles the cigarette beetle (*Lasioderma serricorne*) but is readily identified by the distinctive elongated shape of the last three segments at the end of each antenna, and the wing covers or elytra are striated. Larvae is white in colour and short and stout in shape.

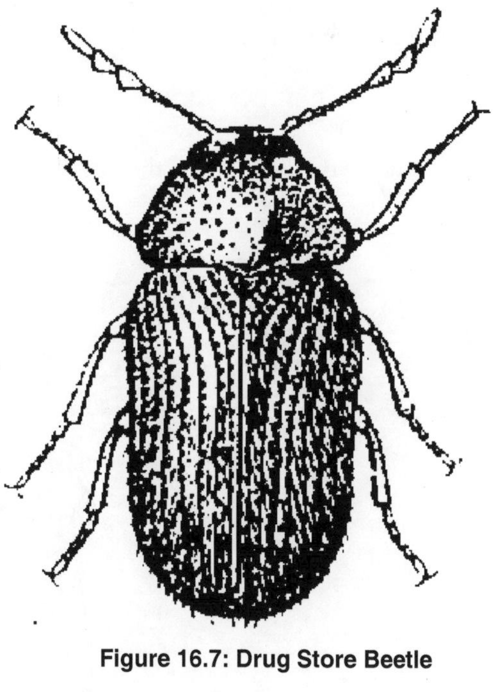

Figure 16.7: Drug Store Beetle

Nature of Damage

The drug store beetle is a cosmopolitan pest that can infest almost any dry animal or plant material. It is particularly noted as a pest of pharmacies where it infests drugs, and domestic premises where it attacks breakfast food, biscuits and herbs, amongst its catholic diet.

Biology

Eggs

The female lays about 75 eggs randomly amongst stored foodstuffs during its life-span of 40 to 90 days.

Larvae

The larva progresses through four to six instars.

Pupae

The last instar larvae constructs a cocoon within which it pupates.

Adult

The adult does not feed and cannot fly.

Warehouse Moths (*Ephestia cautella*)

Order–Lepidoptera, Family–Phycitidae

Marks of Identification

The Adults are short lived and do not feed. They fly well and are active at dawn and dusk. Wings grey with vague darker markings when alive. The larvae (caterpillars) are light pink with a small black spot at the base of each hair. The black pigment at the base of each hair is a key feature in identifying *Ephestia* species.

Nature of Damage

They are important pests of flour mills, food-processing plants and in dried fruit. Silk webbing produced by caterpillar can stop the product flowing freely and make it hard to handle.

Biology

Egg

Female lays 200 eggs on the jute bags. Incubation period is 2 to 17 days.

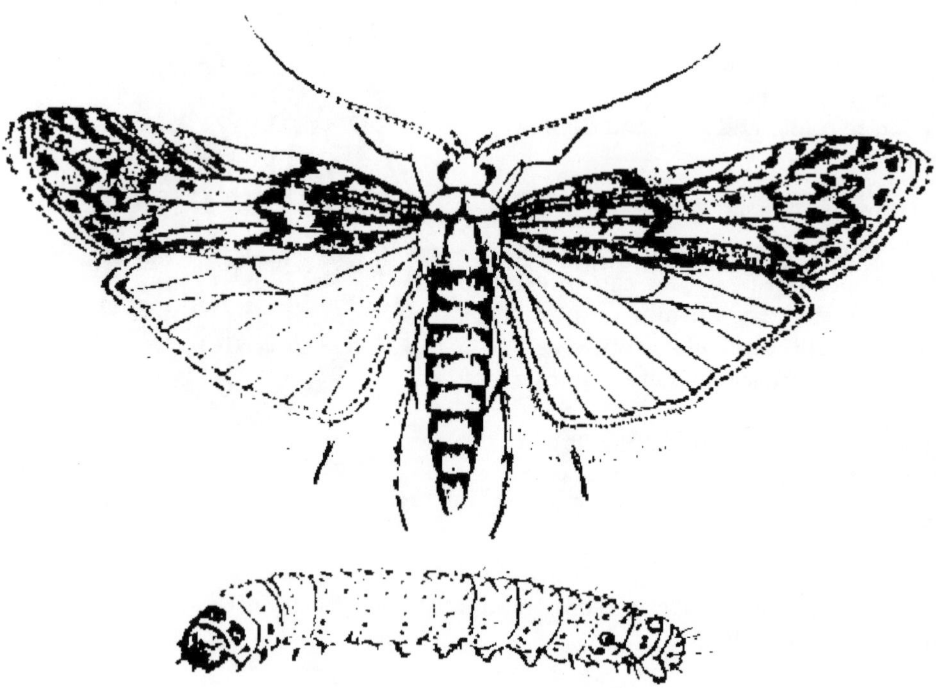

Figure 16.8: Warehouse Moths

Larva

The larvae is grey white in colour. They move freely near the produce, feeding until mature. One larva is observed to consume germ of as many as 64 wheat kernels in its life cycle. The larval period lasts for 20 to 35 days.

Pupa

A silken cocoon for pupation is spun by the mature larvae. The pupal period lasts for 10 days.

Adult

The adult live for 12-14 days. The life cycle completes in 3 to 6 weeks.

Indian Meal Moth (*Plodia interpunctella*)

Order–Lepidoptera, Family–Phycitidae

Marks of Identification

The adult insect has cream and brown wings when the insect is alive. The larvae (caterpillars) are creamy white with no small black spot at the base of each hair.

Nature of Damage

They are important pests of flour mills, food processing plants and in dried fruits. Silk webbing produced by caterpillars can stop the product flowing freely and make it hard to handle.

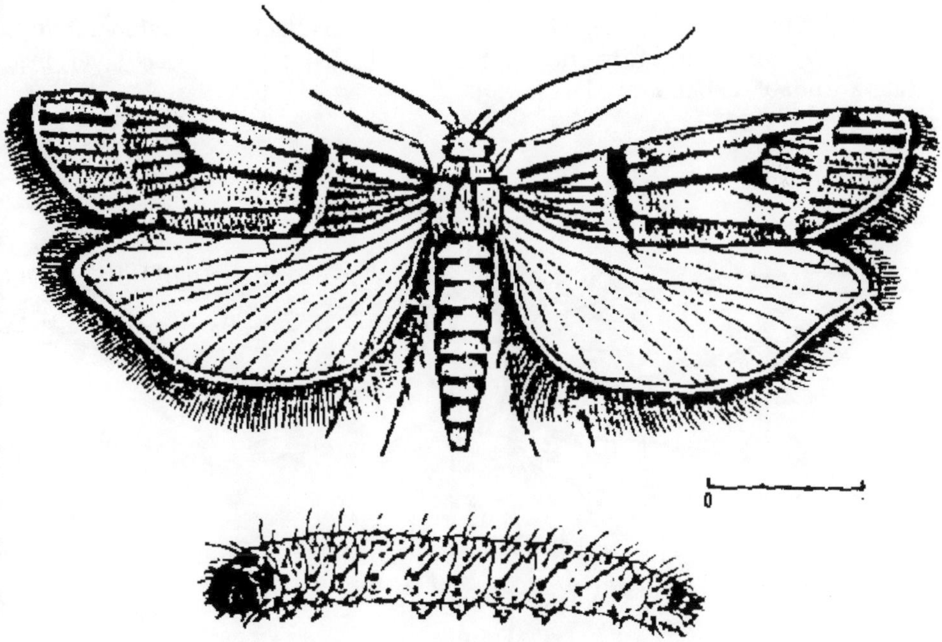

Figure 16.9: Indian Meal Moth

Biology

Egg

Eggs are laid either singly or in group. A single female lays about 200 to 300 eggs.

Larva

The larvae are dirty white in colour. They fed on embryo portion. Larval period lasts for 20 to 35 days.

Pupa

The fully grown larvae web silken cocoon. Pupal period lasts for 10 days.

Adult

The adult moths are pale yellow with red brown wings. The adult live for 12-13 days. The life cycle is completed in 3 to 6 weeks.

Angoumois Grain Moth (*Sitotroga cerealella*)

Order–Lepidoptera, Family–Gelechiidae

Marks of Identification

Wings pale greyish brown with black spot towards tip (in fresh specimens) and are smaller in size than other storage moths.

Nature of Damage

This moth is a pest of whole cereal grain and will attack grain before harvest, particularly maize. Damage to grain in storage only occurs in the surface layer since adults are unable to penetrate deeply.

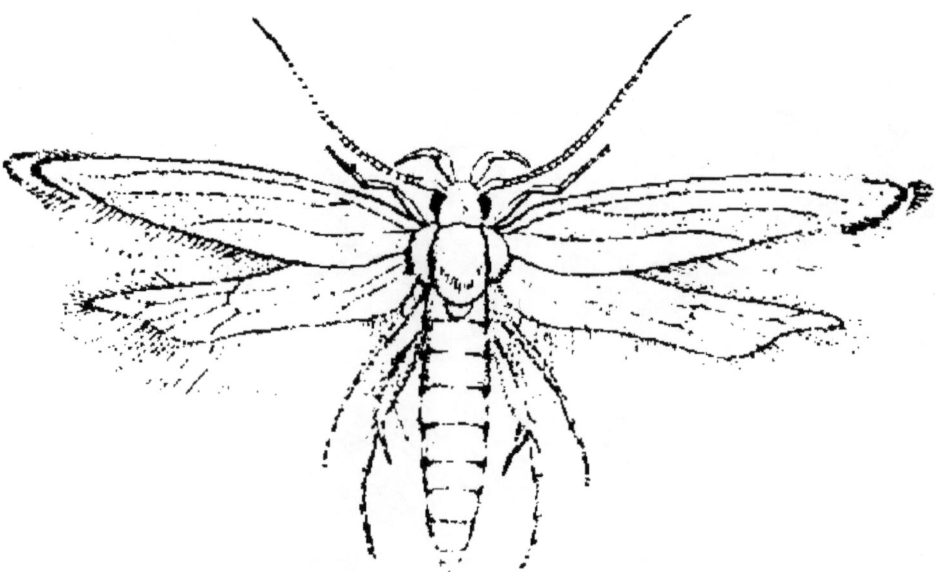

Figure 16.10: Angoumois Grain Moth

Biology

Eggs

Laid on grain surface. Incubation period being 4 to 30 days.

Larvae

Immobile, develop concealed within a single grain. Larva undergoes 4 instars and larval period lasts 12-23 days.

Pupa

Larvae spins silken cocoon inside the grain. Pupae are brown in colour. Pupal period lasts for 7 to 10 days.

Adult

On hatching pupal case often left protruding from grain. Adult moth short lived, does not feed, flies. There are 4-5 generations in a year.

Rice Moth (*Corcyra cephalonica*)

Order–Lepidoptera, Family–Galleridae

Marks of Identifications

The adult moth is about 15 mm long and pale yellowish brown in colour, with faint dark lines along the length of the wings. The young larvae is creamy white in colour.

Nature of Damage

Larvae is only responsible for damage. It pollutes foodgrains with frass, moults and dense webbing. In case of whole grains, kernels are bound into lumps upto 2 kg.

Biology

Eggs

Laid on gunny bags, walls and among grains. A female lays 100 to 200 eggs. Incubation period is 7 days.

Larvae

The young larvae feeds on broken grains. The larval period is 21 to 60 days.

Pupa

Fully mature larvae pupate in silken cocoon in grain. The cocoon is white. Pupal period lasts for about 10 days.

Adult

Adult is biggest among foodgrains infesting moths. Antennae do not cross over wings while insect is at rest. Adult lives for 7-15 days.

Management of Storage Pests

For safe storage of food grains both the preventive and curative measures are required:

(*a*) Preventive measures

"Prevention is better than cure" a common proverb also implies here for management of insect pest of foodgrains. Preventive measures recommend are as follows:

1. Hygiene or sanitation and proper stacking
2. Disinfestations of storage containers/structures/stores.
3. Legal method by imposition of Plan Quarantine Regulations under Destructive Insects Pests Act 1914.

(*b*) Curative Measures

The infestation of stored grain insect pest can be controlled by using both non chemical and chemical control measures.

(c) Non-chemical Control Measures

The non-chemical control measures are ecological, mechanical, physical, cultural, biological, botanical and engineering one.

(*d*) Chemical Control Measures

This method is most popular and effective one. These chemicals may be used for both prophylactic and curative treatments, which include grain protectants, fumigants and pesticide for the control of insects.

Sl.No.	*Name of the Insecticide*	*Concentration of Sprays %*	*Preparation and Dosage*	*Frequency of Treatment*
1.	Malathion 50% EC	0.50%	1:100 at the rate of 3 lit/100 m^2	15 days
2.	Pirimiphosmethyl 50% EC (Actellic)	0.50%	1:100 at the rate of 3 lit/100 m^2	15 days
3.	Pyrethrum WTH 2.0% Pyrethrin EC	0.02%	1:100 at the rate of 3 lit/100 m^2	15 days
4.	Deltamethrin 2.5% W.P.	0.10%	40 gm/lit 3 lit/100 m^2	90 days

Current losses of about 30 per cent are apparently occurring throughout large areas of world. Prevention of these losses would result in:

1. More food for consumption by the farmers
2. More food available for farmers to sell
3. Higher living standards for farmers
4. More food available for non-farming population
5. Higher quality and competitiveness of export commodities in world trade
6. Sounder economy for the country and improvement of its International standing.

Five golden rules for safe storage of stored grains are:

1. Dry and clean the gain before storage.
2. Use dunnage for grain stored in bags.
3. Use modern storage structures or improve the traditional one.
4. Undertake preventive measures with Malathion/Pirimiphos methyl/ Deltamethrin and curative measures with recommended fumigant for insect control.
5. Use recommended rodenticides for rat control in houses and fields.

Index